Mark Sperry

QS–9000 Implementation and Registration

QUALITY AND RELIABILITY

A Series Edited by

1. Designing for Minimal Maintenance Expense: The Practical Application of Reliability and Maintainability, *Marvin A. Moss*
2. Quality Control for Profit: Second Edition, Revised and Expanded, *Ronald H. Lester, Norbert L. Enrick, and Harry E. Mottley, Jr.*
3. QCPAC: Statistical Quality Control on the IBM PC, *Steven M. Zimmerman and Leo M. Conrad*
4. Quality by Experimental Design, *Thomas B. Barker*
5. Applications of Quality Control in the Service Industry, *A. C. Rosander*
6. Integrated Product Testing and Evaluating: A Systems Approach to Improve Reliability and Quality, Revised Edition, *Harold L. Gilmore and Herbert C. Schwartz*
7. Quality Management Handbook, *edited by Loren Walsh, Ralph Wurster, and Raymond J. Kimber*

8. Statistical Process Control: A Guide for Implementation, *Roger W. Berger and Thomas Hart*
9. Quality Circles: Selected Readings, *edited by Roger W. Berger and David L. Shores*
10. Quality and Productivity for Bankers and Financial Managers, *William J. Latzko*
11. Poor-Quality Cost, *H. James Harrington*
12. Human Resources Management, *edited by Jill P. Kern, John J. Riley, and Louis N. Jones*
13. The Good and the Bad News About Quality, *Edward M. Schrock and Henry L. Lefevre*
14. Engineering Design for Producibility and Reliability, *John W. Priest*
15. Statistical Process Control in Automated Manufacturing, *J. Bert Keats and Norma Faris Hubele*
16. Automated Inspection and Quality Assurance, *Stanley L. Robinson and Richard K. Miller*
17. Defect Prevention: Use of Simple Statistical Tools, *Victor E. Kane*
18. Defect Prevention: Use of Simple Statistical Tools, Solutions Manual, *Victor E. Kane*
19. Purchasing and Quality, *Max McRobb*
20. Specification Writing and Management, *Max McRobb*
21. Quality Function Deployment: A Practitioner's Approach, *James L. Bossert*
22. The Quality Promise, *Lester Jay Wollschlaeger*
23. Statistical Process Control in Manufacturing, *edited by J. Bert Keats and Douglas C. Montgomery*
24. Total Manufacturing Assurance, *Douglas C. Brauer and John Cesarone*
25. Deming's 14 Points Applied to Services, *A. C. Rosander*
26. Evaluation and Control of Measurements, *John Mandel*
27. Achieving Excellence in Business: A Practical Guide to the Total Quality Transformation Process, *Kenneth E. Ebel*
28. Statistical Methods for the Process Industries, *William H. McNeese and Robert A. Klein*
29. Quality Engineering Handbook, *edited by Thomas Pyzdek and Roger W. Berger*
30. Managing for World-Class Quality: A Primer for Executives and Managers, *Edwin S. Shecter*
31. A Leader's Journey to Quality, *Dana M. Cound*
32. ISO 9000: Preparing for Registration, *James L. Lamprecht*
33. Statistical Problem Solving, *Wendell E. Carr*
34. Quality Control for Profit: Gaining the Competitive Edge. Third Edition, Revised and Expanded, *Ronald H. Lester, Norbert L. Enrick, and Harry E. Mottley, Jr.*
35. Probability and Its Applications for Engineers, *David H. Evans*
36. An Introduction to Quality Control for the Apparel Industry, *Pradip V. Mehta*
37. Total Engineering Quality Management, *Ronald J. Cottman*
38. Ensuring Software Reliability, *Ann Marie Neufelder*

39. Guidelines for Laboratory Quality Auditing, *Donald C. Singer and Ronald P. Upton*
40. Implementing the ISO 9000 Series, *James L. Lamprecht*
41. Reliability Improvement with Design of Experiments, *Lloyd W. Condra*
42. The Next Phase of Total Quality Management: TQM II and the Focus on Profitability, *Robert E. Stein*
43. Quality by Experimental Design: Second Edition, Revised and Expanded, *Thomas B. Barker*
44. Quality Planning, Control, and Improvement in Research and Development, *edited by George W. Roberts*
45. Understanding ISO 9000 and Implementing the Basics to Quality, *D. H. Stamatis*
46. Applying TQM to Product Design and Development, *Marvin A. Moss*
47. Statistical Applications in Process Control, *edited by J. Bert Keats and Douglas C. Montgomery*
48. How to Achieve ISO 9000 Registration Economically and Efficiently, *Gurmeet Naroola and Robert Mac Connell*
49. QS-9000 Implementation and Registration, *Gurmeet Naroola*

ADDITIONAL VOLUMES IN PREPARATION

QS–9000 Implementation and Registration

Gurmeet Naroola, P.E.
Santa Clara, California

Marcel Dekker, Inc. **New York • Basel • Hong Kong**

Library of Congress Cataloging–in–Publication Data

Naroola, Gurmeet.
QS-9000 implementation and registration / Gurmeet Naroola.
p. cm.— (Quality and reliability; 49)
Includes bibliographical references and index.
ISBN 0-8247-9808-2 (hardcover : alk. paper)
1. Automobile industry and trade—United States—Quality control–
–Standards. 2. Automobiles—Parts—Design and construction—Quality
control. I. Title. II. Series
TL278.N37 1997
629.23'4—dc20 96–32729
CIP

The publisher offers discounts on this book when ordered in bulk quantities. For more information, write to Special Sales/Professional Marketing at the address below.

This book is printed on acid-free paper.

MARCEL DEKKER, INC.
270 Madison Avenue, New York, New York 10016

Current printing (last digit):
10 9 8 7 6 5 4 3 2 1

PRINTED IN THE UNITED STATES OF AMERICA

I dedicate this book to my Grandparents. Thank you for the deep roots.

To Mini and Gaurav, wishing you peace, happiness, and prosperity in your married life.

And to Sonu, my younger brother, always remember, one plus one equals eleven.

Special thanks to Dad, Bob Mac Connell, and Mary Greckel for their input. Thanks also to Peter Lake (IASG Chair), Elizabeth Dorenzo (MIT-CAES), Clare Crawford-Mason (CC-M), the AIAG, and the Supplier Quality Task Force for providing valuable information. Hats off to Bill Berry, my technical writer, for the extra creativity and thanks to Lila Harris, my production editor, for a wonderful job.

About the Series

The genesis of modern methods of quality and reliability will be found in a simple memo dated May 16, 1924, in which Walter A. Shewhart proposed the control chart for the analysis of inspection data. This led to a broadening of the concept of inspection from emphasis on detection and correction of defective material to control of quality through analysis and prevention of quality problems. Subsequent concern for product performance in the hands of the user stimulated development of the systems and techniques of reliability. Emphasis on the consumer as the ultimate judge of quality serves as the catalyst to bring about the integration of the methodology of quality with that of reliability. Thus, the innovations that came out of the control chart spawned a philosophy of control of quality and reliability that has come to include not only the methodology of the statistical sciences and engineering, but also the use of appropriate management methods together with various motivational procedures in a concerted effort dedicated to quality improvement.

This series is intended to provide a vehicle to foster interaction of the elements of the modern approach to quality, including statistical applications, quality and reliability engineering, management, and motivational aspects. It is a forum in which the subject matter of these various areas can be brought together to allow for effective integration of appropriate techniques. This will promote the true benefit of each, which can be achieved only through their interaction. In this sense, the whole of quality and reliability is greater than the sum of its parts, as each element augments the others.

The contributors to this series have been encouraged to discuss fundamental concepts as well as methodology, technology, and procedures at the leading edge of the discipline. Thus, new concepts are placed in proper perspective in these evolving disciplines. The series is intended for those in manufacturing, engineering, and marketing and management, as well as the consuming public, all of whom have an interest and stake in the improvement and maintenance of quality and reliability in the products and services that are the lifeblood of the economic system.

The modern approach to quality and reliability concerns excellence: excellence when the product is designed, excellence when the product is made, excellence as the product is used, and excellence throughout its lifetime. But excellence does not result without effort, and products and services of superior quality and reliability require an appropriate combination of statistical, engineering, management, and motivational effort. This effort can be directed for maximum benefit only in light of timely knowledge of approaches and methods that have been developed and are available in these areas of expertise. Within the volumes of this series, the reader will find the means to create, control, correct, and improve quality and reliability in ways that are cost effective, that enhance productivity, and that create a motivational atmosphere that is harmonious and constructive. It is dedicated to that end and to the readers whose study of quality and reliability will lead to greater understanding of their products, their processes, their workplaces, and themselves.

Edward G. Schilling

Preface

QS-9000, developed by Ford, Chrysler, and General Motors, is fast becoming the accepted quality system requirement for the automotive industry, and suppliers worldwide are facing the need to be registered.

This book provides a unique hands-on, step-by-step method for achieving QS-9000 registration efficiently and economically, aimed at improving customer satisfaction and profitability. It carefully models a number of successful registration efforts personally conducted by the author using the unique TAP-PDSA approach.

QS-9000 Implementation and Registration is an extremely user-friendly book. The chapters are organized sequentially to take the reader from the beginning of a registration effort to "life after registration."

Chapter 1, "The Engine," introduces the user to a unique and proven method (TAP-PDSA) to achieve registration.

Chapter 2, "Behind the Wheel," discusses the "must-know" QS/ISO 9000 knowledge required before a registration effort is started.

Chapter 3, "Put It in Gear," establishes the framework required to launch a successful registration effort. It stresses that the primary goals are to improve the quality system. Included in this chapter are Deming's 14 points of management graphically illustrated by Pulitzer award winning cartoonist Pat Oliphant.

Chapter 4, "Step on the Gas," covers the first few important activities required to launch the registration project.

Chapter 5, "The Drive," explains in detail the Training, Auditing, and Planning requirements as they relate to a QS/ISO 9000 registration effort. A detailed QS-9000 "TAP-PDSA" registration plan is included.

Chapter 6, "Driving Between the Lines," explains the importance and methods of measurements in relation to the QS 9000 project and the quality system.

Chapter 7, "Registrar Selection," addresses the functions of a registrar, the importance of early registrar selection, and the process of selection using the quality, cost, and delivery criteria.

Chapter 8, "Owner's Manual," describes the quality system design and documentation processes.

Chapter 9, "On the Road Again," discusses life after registration. Dr. Deming's 14 points of management are explained in detail.

The primary intended audiences are QS-9000 Project Leads, the Project Team, Quality Management, Manufacturing and Operations Management, and Materials Management. This book can also be effectively used by consultants, educators, trainers, institutions, and professional societies.

I hope the reader will find the information contained here of value. Your suggestions and comments are very welcome.

Gurmeet Naroola

Contents

List of Figures

List of Tables

List of Abbreviations

4Ms:	Man, Machine, Material, and Method
ABC:	Activity-Based Costing
AEC:	Automotive Industry Council
AFNOR:	Association Francaise de Normalisation
AIAG:	Automotive Industry Action Group
ANSI:	American National Standards Institute
APQP:	Advanced Product Quality Planning
AQL:	Average Outgoing Quality Level
ASQC:	American Society of Quality Control
BSI:	British Standards Institute
BVQI:	Bureau Veritas Quality International
CP:	Control Plan
Cpk:	Process Capability Indices
DIS:	Draft International Standard
DnV:	Det Norske Veritas
E:	Environment
EN:	European Norm
EU:	European Union

FMEA: Failure Mode Evaluation and Analysis
IEC: International Electrotechnical Commission
JIT: Just-in-Time
MIL-Q: Military Quality
MIS: Management Information Systems
MIT: Massachusetts Institute of Technology
MR: Management Representative
MRP: Material Requirement Planning
NAA: Needs Assessment Audit
NNI: Netherlands Normalistatie Instituut
PDSA: Plan, Do, Study, Act
PPAP: Production Parts Approval Process
QA: Quality Assurance
QCD: Quality, Cost, and Delivery
QPM: Quality Policy Manual
QSA: Quality System Assessment
QSD: Quality System Documentation
RAB: Registrar Accreditation Board
RvC: RaadVoor de Certificate
SC: Sub-Committee
SCC: Standards Council of Canada
SIC: Standard Industry Code
SPC: Statistical Process Control
SQP: Strategic Quality Improvement Plan
TAG: Technical Action Group
TAG: Technical Advisory Group
TAP: Train, Audit, and Plan
TC: Technical Committee
TLC: Travel and Living Costs
UL: Underwriters Laboratories

QS–9000 Implementation and Registration

Chapter 1
The Engine: TAP-PDSA

Today, the pleasures of driving a new automobile can be short-lived. With rapid advancements in automotive technology, the moment the auto "hits the road" it becomes yesterday's design. Someone around the corner is probably test driving a prototype with tomorrow's features. Competitiveness is becoming fiercer each day and to survive in this age, the need is for innovative quality products at the right cost, ahead of the competition.

Companies feel they are on a racetrack with competitors riding their tail. A single error and a company suddenly finds itself in a pitstop with competition racing ahead. No longer can a company risk failure by reckless driving to the finish line when the development time is becoming shorter and shorter.

The fact is that the pace is breakneck and companies need to give prime importance to planning. Performance by trial and error can be fatal. For this reason, it is important that the process of planning be carefully undertaken so that the results can be *assured.* This planning process is a critical step towards the success of a company

Good planning requires that an honest effort be devoted to training and education in researching the task prior to beginning the project. This process is spelled out in the following section as the "TAP" cycle for establishing a sound plan. The next phase, that of successful implementation of the plan, also involves a sequence of steps. This is described as the "Shewhart" or "PDSA" cycle for implementing the plan. It should be noted that **P**(Plan) is the key common element in both the TAP and PDSA cycles. Details concerning both cycles presented here help achieve successful results in a timely, efficient, and therefore less costly manner.

1 Planning Cycle: TAP

The TAP cycle (Figure 1.1) has three essential elements:

- Train
- Analyze
- Plan

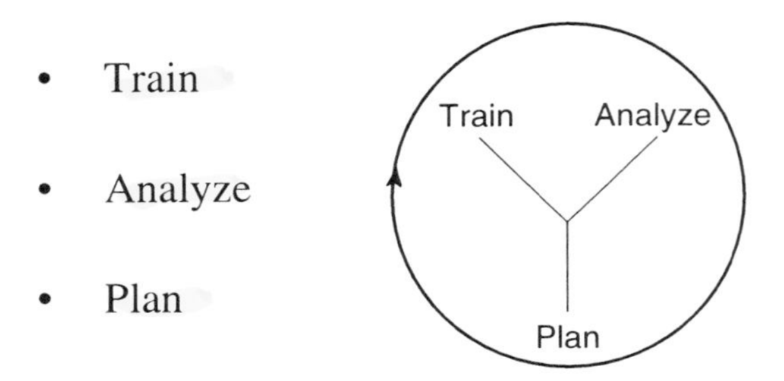

Figure 1.1 TAP Cycle

The TAP cycle requires that the group prepare itself through **training and education** to better understand and predict future customer requirements, then perform an **analysis** of the existing situation before developing an overall implementation **plan** which has the highest probability of success. Finally, a plan is made which is detailed and transparent to all concerned.

2 Implementation Cycle: PDSA

The implementation cycle (Figure 1.2) is shown as the Shewhart PDSA cycle.

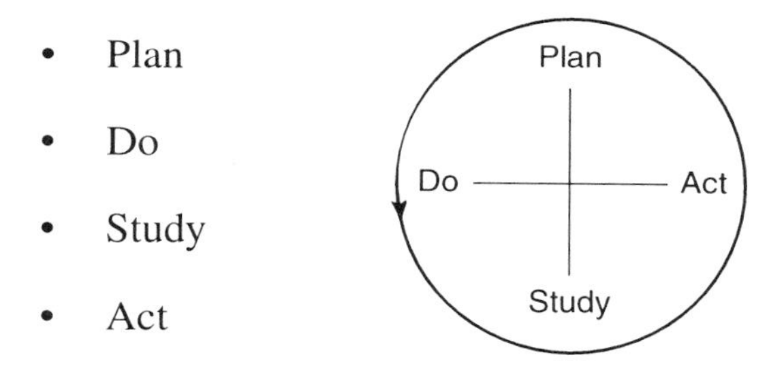

Figure 1.2 Shewhart PDSA Cycle

This cycle explains that the **plan** is implemented in the sequence of first **do**ing what is detailed by the plan, then **study**ing the effectiveness of the plan, and finally **act**ing on the recommended changes to the plan. Thus the steps: Plan, Do, Study, Act.

3 Continuous Improvement Cycle: TAP-PDSA

The two cycles (TAP and PDSA) are illustrated in Figure 1.3. The top cycle is the "driver" and results in proper planning. The bottom cycle is the "driven" and this cycle implements the plan. The diagram also shows plan as a common element to both cycles. Together the TAP and PDSA cycles become very powerful tools for improvement that continuously revolve and drive the results higher and higher.

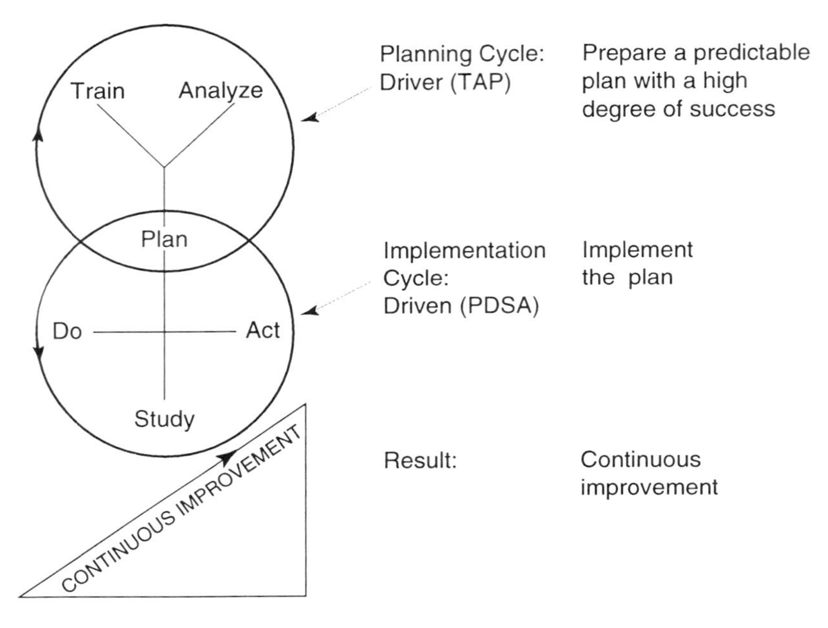

Figure 1.3 TAP-PDSA

3.1 Train

Training is the first step in the TAP cycle. It involves the acquisition of knowledge regarding project matters so that a plan that has a higher probability of success can be predicted. Train to understand the project: What is the purpose and objective? What is the scope of the project? What are the requirements (cost, time, etc.)?

There are many additional benefits of training. Everyone learns to speak the same language and focuses in the same direction, creating a sound starting platform. Training and education also create a body of highly trained employees who are the most important resources of the organization.

3.2 Analyze

This step not only provides the status quo, but also presents a clearer picture of the goals the group wishes to accomplish. Perform an analysis of the existing situation and identify current strengths, weaknesses, and problems. Predict the future requirements. What direction are we headed towards in the future?

3.3 Plan

The better the training, the clearer the analysis, the better the plan. The planning stage may start with a choice between several suggestions. Which direction should we go? Who leads the project? Compare the possible outcomes of the possible choices. Of the several suggestions, which one appears to be most promising in terms of quality, cost, and delivery? The goal is to use the plan that has the highest probability of success and will require minimal changes later during the study step. This plan involves a detailed road map to get from the present condition to the target.

3.4 Do

The do step implements the plan developed by the planning cycle.

3.5 Study

Part of the plan should provide measurement criteria to continuously monitor the project progress. This activity is performed at the study stage. Compare the results with the plan. Are we progressing according to the plan? Are the results acceptable? Does the plan require any modification?

3.6 Act

The outcome of the study stage dictates this step. If all is well, the implementation moves on. If the outcome did not provide the desired results, appropriate changes are made to the plan and the cycles are repeated. This leads back to the plan, which is continuously improved through ongoing training and analysis. This is like one gear driving another gear with TAP as the driver gear and PDSA as the driven gear.

A typical TAP-PDSA continuous improvement worksheet is shown in Figure 1.4.

4 TAP-PDSA and QS-9000

TAP-PDSA can be used in any project successfully to produce outstanding results. The QS-9000 registration effort is no exception.

The QS-9000 registration project kicks off with **training and education,** making sure that all involved are well trained in the various QS-9000 requirements, talk the same language, that of QS-9000, and look in the same direction; again that of QS-9000 registration.

Upon having a clear understanding of requirements, a detailed **analysis** of the existing quality system is performed against the QS-9000 requirements. This step helps the status quo, weaknesses, and strengths of the quality system. It also helps to properly determine the resource needs (i.e., cost, manpower, etc.) to successfully achieve the target of registration (customer satisfaction and profitability). This analysis usually takes place in the form of the Needs Assessment Audit.

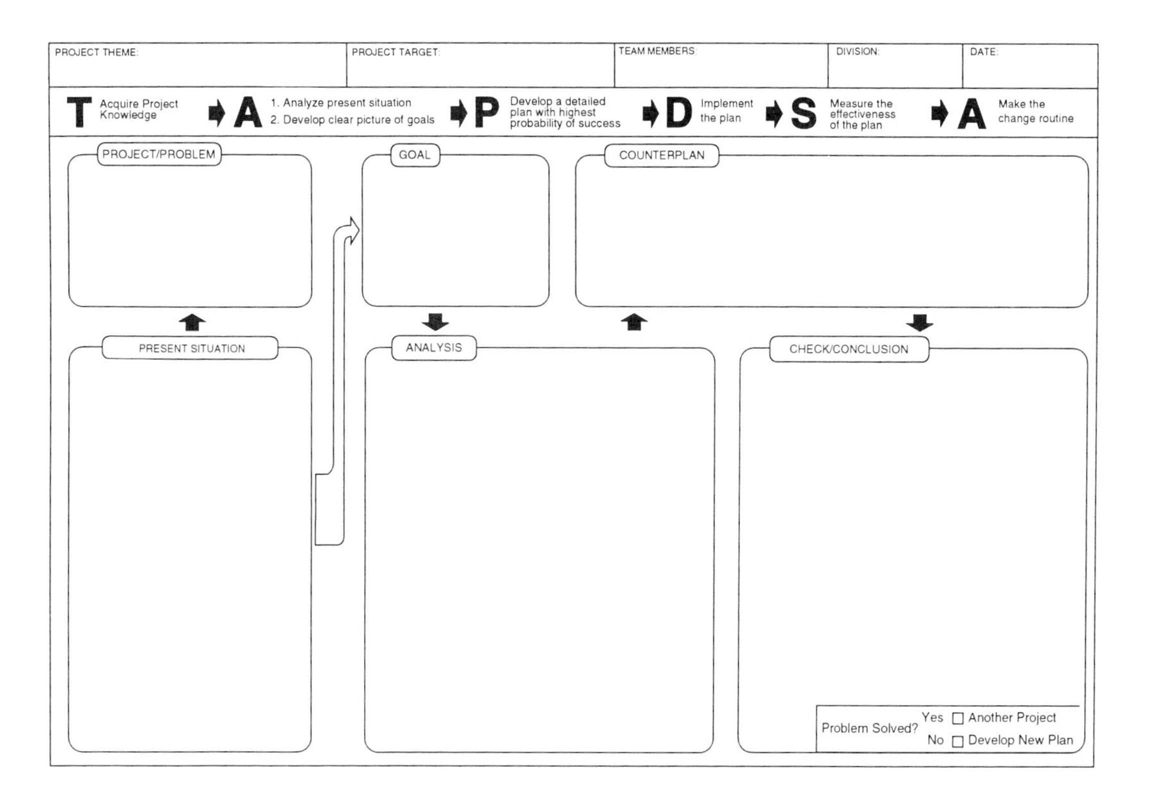

Figure 1.4 TAP-PDSA Continuous Improvement Worksheet

With a clear understanding of the present situation, target, and the resources required, a detailed implementation **plan** with a higher success rate can be predicted. This plan is then implemented through the Do, Study, and Act steps of the Shewhart cycle.

The benefits of this approach are many. The principal ones are:

- A superior quality system will be developed.
- Registration will be achieved sooner.
- Registration costs will be significantly lower.

Note: Chapter 5 provides detailed information on the training, auditing, and planning requirements for the QS-9000 registration effort.

5 Conclusion

Implementation problems are a direct result of poor planning. Invest adequate resources in the planning cycle. It will result in a shorter and better implementation of the QS-9000 system.

Chapter 2
Behind the Wheel

This chapter answers in detail the "must-know questions" most frequently asked prior to beginning a QS-9000/ISO 9000 registration project. This information provides the reader with a basic understanding regarding QS and ISO organizations, the QS-9000 quality system requirements and ISO 9000 standards, and the registration process. Also included are parts of the International Automotive Sector Group (IASG) sanctioned interpretations. This information establishes a strong foundation for registration.

The questions and answers are grouped by topic:

The Organizations

- Supplier Quality Requirement Task Force
- International Automotive Sector Group (IASG)

- Automotive Industry Action Group (AIAG)
- ISO
- ISO/TC-176

The Requirements/Standards

- QS-9000 and ISO 9000
- QS-9000 Supplemental Requirements
- QS-9000 Reference Manuals
- ISO 9000 Guidance Standards

The Registration Process

- Registrars
- Registration Process

IASG Sanctioned Interpretations (partial listing)

1 The Organizations

1.1 Supplier Quality Requirement Task Force

1.1.1 What Is the Supplier Quality Requirement Task Force?

The Supplier Quality Requirement Task Force is an international ad hoc working group established in June 1988 by the Purchasing and Supply Vice Presidents of Chrysler, Ford, and General Motors and consists of representatives from these companies.

1.1.2 What Is the Purpose of the Supplier Quality Requirement Task Force?

The purpose of the Supplier Quality Requirement Task Force is to harmonize and promote supplier quality system requirements. The group meets regularly at the AIAG and addresses issues regarding the seven major deliverables:

- QS-9000
- APQP and Control plans
- PPAP
- QSA
- MSA
- SPC
- FMEA

In addition the group discusses related topics such as harmonization of automotive standards, spreading QS-9000 globally, and approving and issuing interpretations.

The meetings include representatives from American Society for Quality Control (ASQC) and the AIAG acting as observers.

1.1.3 How Can the Supplier Quality Requirement Task Force Be Contacted?

The following members of the Supplier Quality Task Force can be contacted for suggestions and improvements to the QS-9000 Quality System Requirements.

Chrysler	Ford	General Motors
Norrid Jr. Warren	Steve Walsh	R. Dan Reid
800 Chrysler Drive East	17101 Rotunda	6060 W. Bristol Road
Auburn Hills, MI 48236	Dearborn, MI 48121	Flint, MI 48554
Phone: 810-576-2701	Phone: 313-845-8442	Phone: 810-857-0295

1.2 International Automotive Sector Group

1.2.1 What Is the International Automotive Sector Group?

The International Automotive Sector Group (IASG) is an international ad hoc working group consisting of representatives from:

- Big Three Recognized Accreditation Bodies
- QS-9000 Qualified Registrars
- Representatives of the Chrysler/Ford/General Motors Suppliers Supplier Quality Requirement Task Force
- Tier 1 Automotive Suppliers

1.2.2 What Is the Purpose of the Group?

The group meets regularly to discuss and resolve interpretation issues relative to the QS-9000 criteria and third party registration of auto suppliers to QS-9000. It publishes the "IASG Sanctioned QS-9000 Interpretations" which are sanctioned and recognized by the Chrysler, Ford, General Motors Supplier Quality Requirement Task Force, the participating ISO 9000 accreditation bodies, and QS-9000 qualified registrars.

These interpretations are periodically revised, and updated interpretations are released for all interested parties.

1.2.3 How Can the IASG Be Contacted?

To submit questions or issues to the IASG for consideration, fax inquiries, in English, to:

IASG Fax Voice Mail Box: 614-847-8556

1.3 Automotive Industry Action Group

1.3.1 What Is the Automotive Industry Action Group?

The Automotive Industry Action Group (AIAG) is a not-for-profit trade association dedicated to providing a forum to encourage communication, standardize business practices, and provide education for North American vehicle manufacturers and suppliers.

AIAG represents the North American Automotive industry with a major voice in setting standards in cooperation with such organizations such as the American National Standards Institute, the Federation of Automated Coding Technologies, and Electronic Data Interchange (EDI) for Administration, Commerce and Trade.

AIAG also devotes significant resources to developing education and training materials and programs. AIAG offers:

- Seminars, videotapes, and publications on a full spectrum of automotive subjects including QS-9000 and the related reference manuals (PPAP, APQP, MSA, SPC, FMEA, etc.).
- Training in Activity Based Cost Management, EDI, NAFTA, and Content Reporting.

AIAG also holds an annual Auto-Tech conference designed to help automotive manufacturers and suppliers understand and implement new technology applications in their own companies.

1.3.2 AIAG's Mission

AIAG seeks to improve the global productivity of its members and the North American Automotive Industry by providing an organization to:

- Foster cooperation and communication between trading partners to improve and reduce variation in business processes and practices.
- Address existing and emerging common issues and apply new and current technology to increase the efficiency of the industry.
- Promote a sense of urgency in adopting developed business practices.
- Cooperate and communicate with other industry, governmental and technical organizations.

1.3.3 Where is AIAG Located?

AIAG
26200 Lahser Road, Suite 200
Southfield, MI, 48034
Phone: 810-358-3003 (Order Desk) or
810-358-3253

1.4 ISO

1.4.1 What Is ISO?

ISO is a worldwide federation, headquartered in Geneva, Switzerland, with over 100 member countries. It develops standards for all industries except those industries related to the electric and electronics disciplines, which are the venue of the International Electrotechnical Commission (IEC). Together, these two groups form the largest and most comprehensive world-wide, nongovernmental forums for voluntary industrial and technical collaboration at the international level.

1.4.2 What Does ISO Mean?

ISO doesn't stand for anything, although it functions as an acronym when referring to the Geneva-based International Organization for Standardization. According to ISO officials, the organization's short name was borrowed from the Greek word *isos*, meaning "equal." *Isos* also is the root of the prefix "iso," which appears in "isometric" (of equal measure or dimension) and "isonomy" (equality of laws or of people before the law). Its election was based on the conceptual path taken from "equal" to "uniform" to "standard."

1.4.3 Who Were the Founding Members of ISO?

The founding members consisted of fourteen industrialized countries from Europe, the United States, and the British Commonwealth. The American National Standards Institute (ANSI) was the founding member representative for the United States.

1.4.4 Where Is ISO Located?

The headquarters secretariat for the International Organization for Standardization is located in Geneva, Switzerland.

Mailing address:

1, rue de Varmbe
Caste postale 56
CH1211 Geneve/Suisse

Phone: 41 22 749 0111
Telex: 41 22 733 3430
Internet: CENTRAL@ISOCS.ISO.CH

1.4.5 How Does ISO Operate?

ISO work is decentralized, carried out by over 180 active technical committees (TCs) and over 620 active sub-committees (SCs). They are supported by technical secretariats in thirty-four countries. The General Secretariat in Geneva assists in coordinating ISO operations world-wide. Over 30,000 specialists in their fields develop international standards. These specialists are nominated by ISO members to participate in committee meetings, and to represent the consolidated views and interests of industry, government, labor, consumers, and other interested parties in the standards development process. The actual standards preparation is done by ISO Technical Committees (ISO/TCs) assembled and assigned responsibility for developing related standards. The working technical committees and sub-committees meet periodically in world-wide locations to develop or review standards work. TCs are assisted by member country Technical Action Groups (TAG) to provide counsel for standards under development or revision. Requests to develop new standards come from member countries to ISO, which evaluates the need and benefit before a TC is assembled.

1.5 ISO/TC-176

1.5.1 What Is ISO/TC-176?

Technical work in the International Organization for Standardization is handled by technical committees (TCs). ISO/TC-176 is the 176th ISO technical committee and was formed in 1979 to address the needs for standards in Quality Management and Quality Assurance. ISO/TC-176 completed the development of the ISO 9000 core series of standards in 1987.

ISO technical committees are numbered serially as they are established, starting with TC-1 (screw threads) in 1946 to TC-207 (environmental management) in 1993. A number is never repeated, nor is a technical committee assigned a new project when its work is done.

1.5.2 How Is TC-176 Organized?

TC-176 consists of a central committee and three sub-committees (SCs) to deal with specific areas (Figure 2.1).

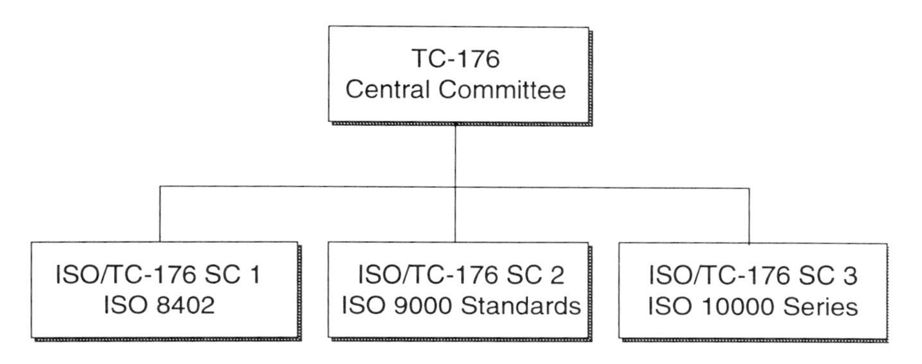

Figure 2.1 TC-176 Organization

ISO/TC-176 SC 1

SC 1, responsible for the development of concepts and terminology, created ISO 8402. The Secretariat is Association Francaise de Normalisation (AFNOR), France.

ISO/TC-176 SC 2

SC 2 was responsible for the development of quality systems guidelines and standards (ISO 9000 through 9004). The Secretariat is British Standards Institute (BSI), United Kingdom.

ISO/TC-176 SC 3

SC 3 was responsible for the development of supporting technology guidelines (ISO 100XX series). The Secretariat is Netherlands Normalisatie Instituut (NNI), Netherlands.

The five national associations participating in ISO/TC-176 as organizers are BSI, ANSI, SCC, AFNOR, and NNI.

2 The Standards

2.1 QS-9000 and ISO 9000

2.1.1 What Are QS-9000 Requirements and ISO 9000 Standards?

QS-9000 is the quality system requirement of Ford, Chrysler, General Motors, truck manufacturers, and other subscribing companies. It defines their quality system expectations.

The ISO 9000 series are a set of generic standards that state the requirements for an acceptable quality management system.

2.1.2 What Is the Structure of QS-9000 and ISO 9000?

The QS-9000 quality system requirement is organized into three sections:

- ISO 9000-Based Requirements
- Sector-Specific Requirements
- Customer-Specific Requirements

Section I: ISO 9000-Based Requirements

Section I is known as the ISO 9000-based requirement and contains the ISO 9001 requirements in their entirety. Also included in this section are additional requirements of the big three automotive companies. The ISO 9001 requirements are in italic while additional requirements are in roman type.

Section II: Sector-Specific Requirements

This section contains sector-specific requirements and addresses the following three topics:

- Production Parts Approval Process
- Continuous Improvement
- Manufacturing Capabilities

Section III: Customer-Specific Requirements

This section addresses the customer-specific requirements for the big three and truck manufacturers, such as third party registration requirements, special characteristics, identification symbols, etc. The overall QS-9000 structure is shown in Figure 2.2.

The ISO 9000 standards are of two types:

- Conformance
  - ISO 9001
  - ISO 9002
  - ISO 9003
- Guidance
  - ISO 9000 is a guidance standard for selecting the proper conformance standard.
  - ISO 9004 is a guideline for quality management and quality system elements.

As the standards and guidelines have become more widespread, the guidelines have been expanded to cover more and more industry and service situations. For example, ISO 9000 has become ISO 9000-1,

9000-2, 9000-3, and 9000-4, and ISO 9004 has become ISO 9004-1, 9004-2, 9004-3, 9004-4, 9004-5, and 9004-7. (See Section 6.2, "What Guidance Standards Are Available," for details.) Figure 2.3 shows the structure of the ISO 9000 series of standards.

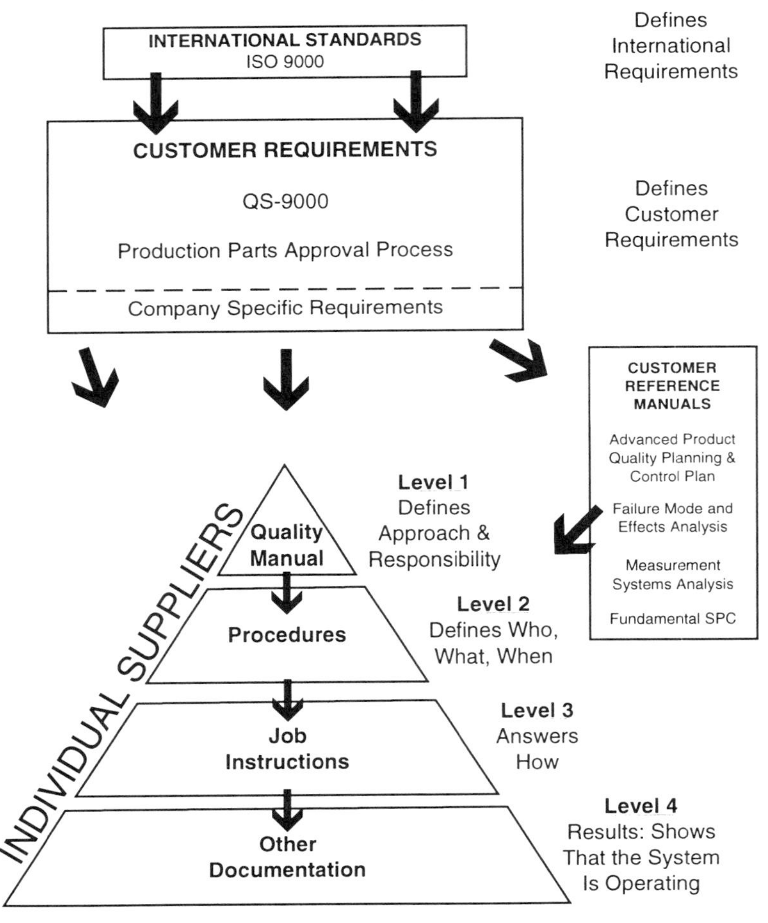

Source: Reprinted from the Advanced Product Quality Planning and Control Plan Manual with permission from the Supplier Quality Task Force

Figure 2.2 QS-9000 Structure

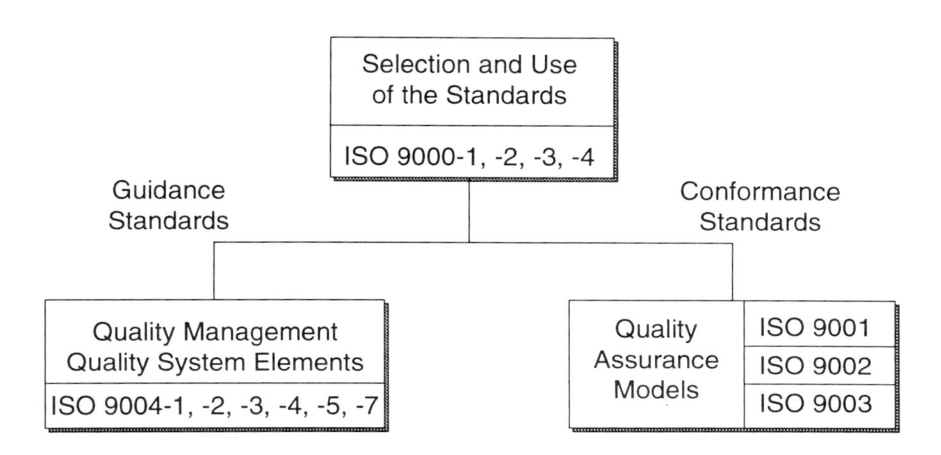

Figure 2.3 The Structure of ISO 9000 Standards

2.1.3 Why Were QS-9000 and ISO 9000 Developed?

QS-9000 was developed by the big three supplier quality group to harmonize their quality requirements (Table 2.1).

Table 2.1: Harmonization of Quality System Requirements

Company	Before	Today
Ford	Q-101	QS-9000
Chrysler	SQA	QS-9000
GM	Targets for Excellence	QS-9000
Truck Mfg.	N/A	QS-9000

The initial need for the ISO 9000 International Quality System Standards was for two-party contractual situations, between customer and supplier. The objective was to increase customer confidence in the quality systems of its suppliers. With common standards, the approach also served to make supplier quality systems more uniform. As manufacturers moved from vertical integration of manufacturing processes to increased dependence upon suppliers and sub-contractors, monitoring the quality of product through source inspections and supplier quality audits became more important.

The ISO 9000 standards are unique standards for quality systems rather than specification or product standards. The predecessors of the development of the International Quality Standard were national standards such as British Standard BS 5750 (a source document for the ISO 9000 series of quality standards), military standards (i.e., MIL-Q-9858A) in the United States, and a variety of industry standards and even company standards for inspection and quality assurance.

2.1.4 What Is the Scope of QS-9000 and ISO 9000?

QS-9000 applies to all internal and external suppliers of: a) production materials, b) production or service parts, c) heat treating, painting, plating, or other finishing services directly to Chrysler, Ford, General Motors, or other OEM customers subscribing to QS-9000.

ISO 9000 standards are in use in every industrialized country and most emerging countries in the world. They are rapidly becoming the de facto global quality assurance standards for industry, business, education, and government. One of the achievements of the writers of the International Standards is that they are completely generic and can be applied to almost any situation where a quality system is needed.

2.1.5 How Do QS-9000 Requirements and ISO 9000 Standards Relate to Each Other?

QS-9000 is the most comprehensive standard and contains 23 requirements (Table 2.2), whereas ISO 9001 contains 20 requirements and ISO 9002 contains 19 requirements.

Table 2.2: Quality System Requirements

Quality System Requirements		9001	9002	QS-9000
1.	Management Responsibility	X	X	X
2.	Quality System	X	X	X
3.	Contract Review	X	X	X
4.	Design Control	X		X
5.	Document and Data Control	X	X	X
6.	Purchasing	X	X	X
7.	Control of Customer-Supplied Product	X	X	X
8.	Product Identification and Traceability	X	X	X
9.	Process Control	X	X	X
10.	Inspection and Testing	X	X	X
11.	Control of Inspection, Measuring, and Test Equipment	X	X	X
12.	Inspection and Test Status	X	X	X
13.	Control of Non-Conforming Product	X	X	X
14.	Corrective and Preventive Action	X	X	X
15.	Handling, Storage, Packaging, Preservation, and Delivery	X	X	X
16.	Control of Quality Records	X	X	X
17.	Internal Quality Audits	X	X	X
18.	Training	X	X	X
19.	Servicing	X	X	X
20.	Statistical Techniques	X	X	X
21.	Production Parts Approval Process			X
22.	Continuous Improvement			X
23.	Manufacturing Capabilities			X

2.1.6 How Do I Select a Proper Requirement/ Standard for Use?

For QS-9000, verification of conformance to ISO 9001(or ISO 9002 for suppliers that are not responsible for the design of any product supplied to any customer subscribing to this document) is a necessary condition to for registration to QS-9000.

Note: Design responsible suppliers cannot achieve QS-9000 registration at an ISO 9002 level.

ISO 9000-1, Guidelines for the Selection and Use, was written to assist in selection of the proper ISO standard to use. However, there is a very simple rule of thumb:

- ISO 9001 – If you design products
- ISO 9002 – If you manufacture only

2.1.7 How Are QS-9000 and ISO 9000 Interpreted?

In QS-9000 the word "shall" indicates mandatory requirements. The word "should" indicates a preferred approach. Suppliers choosing other approaches must be able to show that their approach meets the intent of QS-9000. Where the words "typical" and "examples" are used, the appropriate alternative for the particular commodity or process should be chosen. Paragraphs marked "note" are for guidance.

Note: Definitions for words not in common use are provided in the Vocabulary Standard: ISO 8402.

In the ISO 9000 conformance standards the word "Shall" without question is most important. "Shall" is a command: "You will do it!" To avoid confusion, the word "will" is not used. Modifiers, such as "on occasion, shall" are not used. Shall tells the readers that something is required to be done and objective evidence of actions performed is needed. You must be able to demonstrate full compliance all the time.

"As necessary," "as appropriate," "may," "normally," "should," "where applicable," "where practicable" – When any of the above terms are used, they provide flexibility of application to an organization and the degree of compliance is subject to the specific organization's interpretation.

2.1.8 How Often Are QS-9000 and ISO 9000 Revised?

The QS-9000 Standard was first released in August 1994. The worldwide demand for QS-9000 created the need for a second revision that was released in February 1995.

The changes are minor in nature. In particular, revisions recommended by the companies' European affiliates have been included to facilitate QS-9000 implementation throughout Europe.

Input received from the Task Force Group indicates that the QS-9000 will be examined for improvements in 1997.

The original series of standards, ISO 9000, 9001, 9002, 9003, and 9004, were first issued in 1987. Subsequent revisions are always listed with the standard (e.g., ISO 9001:1994). It was the intent of ISO to review the standards every five years. The last revision cycle started in 1992 and the revisions were released in 1994. When revisions are at the final stages, they are available as Draft International Standards (DIS). The next major revision is expected to be released in 1999.

2.1.9 Who Has Adopted QS-9000 and ISO 9000?

QS-9000 is fast becoming the de facto automotive industry requirement and has been adopted by the following:

- General Motors Holden Australia
- Ford Australia
- Toyota Australia
- Mitsubishi Australia

On the other hand, by the end of 1995 the ISO 9000 series of International Standards will have been adopted by more than a hundred countries. Table 2.3 shows the ISO 9000 Standards nomenclature for major industrialized countries.

Typically countries that adopt the standard assign a standards name and number consistent with their existing standards. For example, the United States, which first assigned a Q90 series of numbers through the American National Standards Institute (ANSI) and the American Society for Quality Control (ASQC), in the 1994 revisions use the more applicable Q9000 series of numbers. The European Union (EU) has adopted the ISO 9000 series as European Norm (EN) 29000.

2.1.10 Where Can QS-9000 and ISO 9000 Be Obtained?

The QS-9000 requirements are available from AIAG.

Automotive Industry Action Group (AIAG)
26200 Lahser Road, Suite 200
Southfield, MI, 48034
Tel: 810-358-3570
Fax: 810-358-3253

The ISO 9000 Standards are available from ISO in Geneva, Switzerland. However, in the United States it is more convenient to order from:

The American National Standards Institute (ANSI)
11 West 42nd St.
New York, NY 10036
Tel: 212-642-4900
Fax: 212-398-0023

The American Society for Quality Control (ASQC)
611 East Wisconsin Ave.
Milwaukee, WI 53201
Tel: 800-248-1946
Fax: 414-272-1734

Table 2.3: World-Wide Equivalents of ISO 9000 Standards

Country	ISO 9000 Standards Nomenclature
Australia	AS 3900
Belgium	NBN-EN 29000
Brazil	NB 9000
Canada	Z 299
Chile	NCH-ISO9000
China	GB/T 10300
Denmark	DS/ISO 9000
Finland	SFS-ISO9000
France	NF EN 29000
Germany	DIN ISO 9000
Greece	ELOT EN29000
Iceland	IST ISO 9000
India	IS 14000
Ireland	I.S./ISO 9000
Italy	UNI/EN 29000
Japan	JIS Z 9900
Mexico	NOM-CC 2
Netherlands	NEN ISO 9000
Norway	NS ISO 9000
Portugal	EM 29000
Singapore	SS 306
South Africa	SABS 0157
Spain	UNE 66-9000
Sweden	SS-ISO 9000
Switzerland	SN-EN 29000
United Kingdom	BS 5750
United States	Q9000

2.2 QS-9000 Supplemental Requirement

The Automotive Electronics Council (AEC) under the guidance of Chrysler/Ford/General Motors developed the QS-9000 Semiconductor Supplement.

2.2.1 What Is the QS-9000 Semiconductor Supplement?

The supplement defines addition common quality system requirements unique to the producers of semiconductor devices.

2.2.2 How to Use the Semiconductor Supplement

It is recommended that the QS-9000 semiconductor supplement be used together with QS-9000 as one document.

2.3 QS-9000 Reference Manuals

2.3.1 What Are the QS-9000 Reference Manuals?

The Supplier Quality Task Force developed the following reference manuals to supplement the QS-9000 quality system requirement.

- Advanced Product Quality Planning (APQP) and Control Plan
- Potential Failure Mode and Effect Analysis (FMEA)
- Measurement System Analysis (MSA)
- Statistical Process Control (SPC)
- Quality System Assessment (QSA)
- Production Parts Approval Process (PPAP)

APQP and Control Plan Reference Manual

What is the APQP and Control Plan manual?

This manual communicates to suppliers (internal and external) common Product Quality Planning and control plan guidelines developed jointly by Chrysler, Ford, or General Motors. It provides guidelines to enable the supplier to create a product quality plan which will support the development of products that will satisfy the customer.

- APQP

The APQP section provides a matrix showing the various phases a product progresses through, from initial concept design, to approval, prototype build, pilot production, and product launch (Figure 2.4).

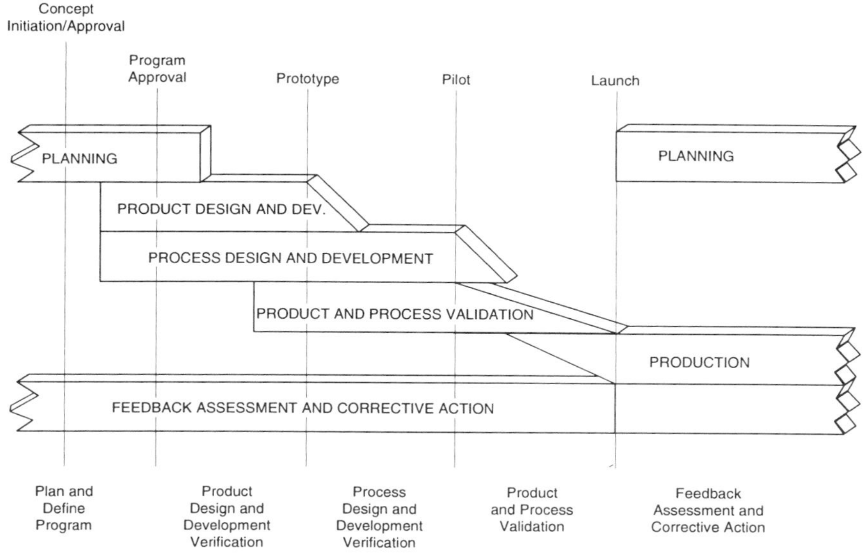

Source: Reprinted from the Advanced Product Quality Planning and Control Plan Manual with permission from the Supplier Quality Task Force

Figure 2.4 Product Quality Planning Timing Chart

- Control Plan

The control plan (CP) section describes a systematic and structured quality planning approach for producing/controlling the product and its processes. It describes the 4M (man, machine, material, and method) operation components utilizing the (5Ws and 1H) question and answer format. A typical example is shown in Figure 2.5.

When are the APQP and CP used?

The APQP and CP are active documents and are used and maintained throughout the product life cycle.

Who should use them?

The APQP and the CP are the responsibility of the engineering, manufacturing, materials, purchasing quality, sales, and field service departments as well as subcontractors and customers, as appropriate.

Potential Failure Mode and Effect Analysis (FMEA) Reference Manual

What is the FMEA?

FEMA is a systematic method or tool intended to recognize, evaluate, and prevent the potential failure modes of a product or process and its effects from reaching the customer.

There are two types of FMEAs, namely:

- Design FMEA
- Process FMEA

Page ___ of ___

☐ Prototype ☐ Pre-launch ☐ Production Control Plan Number	Key Contact/Phone	Date (Orig.)	Date (Rev.)
Part Number/Latest Change Level	Core Team	Customer Engineering Approval/Date (If Req'd)	
Part Number/Description	Supplier/Plant/Approval/Date	Customer Quality Approval/Date (If Req'd)	
Supplier/Plant · Supplier Code	Other Approval/Date (If Req'd)	Other Approval/Date (If Req'd)	

Part/ Process Number	Process Name/ Operation Description	Machine, Device, Jig, Tools for Mfg.	Characteristics			Special Char. Class.	Methods					Reaction Plan
			No.	Product	Process		Product/Process Specification/ Tolerance	Evaluation Measurement Technique	Sample		Control Method	
									Size	Freq.		

Source: Reprinted from the Advanced Product Quality Planning and Control Plan Manual with permission from the Supplier Quality Task Force

Figure 2.5 Control Plan

Design FMEA

The design FMEA (Figure 2.6) is an analytical method used at the product design phase to ensure that the all possible product design failure modes have been determined, evaluated, and addressed to minimize product failures.

It also helps reduce the risk of failures during the manufacturing process. The key is to ensure that the product designed will meet the design intent or customer requirement.

This is a living dynamic document and is usually initiated during the product concept phase, continuously updated throughout the product development phase and completed before product approval.

Prcess FMEA

A process FMEA is also an analytical technique and is used at the process design phase to ensure that all possible process failure causes have been determined, evaluated and addressed.

The process FMEA like the design FMEA is a living document and is usually initiated before the feasibility stage and takes into account all aspects of the manufacturing operation

When should FMEAs be used?

Both the design and process FMEA are dynamic methods and are meant to be utilized continuously during the product or process life-cycle to foster continuous improvement.

The FMEA is a "Do it right the first time" approach and is meant to be a "before the event action" exercise.

Note: The design FMEA is a requirement of the Production Parts Approval Process (PPAP) and must be completed whenever a PPAP submission is required.

POTENTIAL
FAILURE MODE AND EFFECTS ANALYSIS
(DESIGN FMEA)

___ System
___ Subsystem
___ Component ____________

Design Responsibility ____________

FMEA Number ____________

Page ______ of ______

Prepared By ____________

Model Year(s)/Vehicle(s) ____________

Key Date ____________

FMEA Date (Orig.) ____________

Core Team ____________

Item / Function	Potential Failure Mode	Potential Effect(s) of Failure	Sev	Class	Potential Cause(s)/ Mechanism(s) of Failure	Occur	Current Design Controls	Detec	R. P. N.	Recommended Action(s)	Responsibility and Target Completion Date	Action Results				
												Actions Taken	Sev	Occ	Det	R. P. N.

Source: Reprinted from the Advanced Product Quality Planning and Control Plan Manual with permission from the Supplier Quality Task Force

Figure 2.6 Design FMEA (Typical)

Who should use FMEAs?

The FMEA is a team-oriented tool for engineers with expertise in design, manufacturing, assembly, service, quality, and reliability.

The FMEA stimulates the interchange of ideas between the various affected functions in an organization.

Measurement System Analysis (MSA) Reference Manual

What is the MSA reference manual?

This reference manual provides guidance on how to establish and maintain a measurement system. It explains a two-phase approach to preparing and maintaining a capable measurements system that possesses the statistical properties and confidence required by management.

MSA covers the following topics:

- "Gage R & R"
- Accuracy
- Repeatability
- Reproducibility
- Stability
- Linearity

Phase One

Phase one assessment is the testing conducted to determine if the measurement system has the statistical properties needed to perform the required task.

Phase Two

Phase two assessment is the periodic retesting performed to determine if the measurement system has remained acceptable.

When is it used?

The MSA reference manual is used to prepare a system for assessing the quality of measurements primarily associated with the manufacturing processes.

Who uses it?

Anybody responsible for specifying, using, maintaining/calibrating, repairing, and controlling the measurement system is a potential user of this reference manual.

Fundamental Statistical Process Control (SPC) Reference Manual

What is the SPC reference manual?

This reference manual represents the SPC methods used commonly agreed upon by Chrysler, Ford, and General Motors. It is a combination and consolidation of the SPC requirements of all three companies. The SPC reference manual describes several basic statistical methods which can be used effectively for continuously improving processes and products. It covers the following topics:

- Process control background
- Variable control charts (X-bar and R charts, X-bar and s charts, median charts, and X-MR (individuals and moving range)
- Attribute control charts (p chart, np chart, c chart and u chart)
- Measurement system analysis

When is the SPC reference manual used?

The manual should be consulted whenever the application of statistical method is directed by the standard.

Note: Whenever a PPAP submission is required, the process capability results showing conformance to customer requirements for key characteristics, with supporting data such as control charts, must be completed.

Who should use the SPC reference manual?

The SPC reference manual is aimed at practitioners and managers beginning the application of statistical methods. It is first and foremost a training manual.

Quality System Assessment (QSA) Manual

What is the QSA manual and what is its purpose?

The QSA manual is an audit checklist used to determine conformance to the Quality System Requirement, QS-9000. It consists of a series of planned questions that cover the entire quality system. There is a scoring section that helps determine the system compliance and effectiveness level. This is an excellent tool for the needs assessment audit.

Proper use of the QSA will promote consistency between activities and personnel using QS-9000. The QSA process is defined in Appendix A of QS-9000.

When is it used?

The QSA can be used in many different ways:

- As self-assessment
- By a supplier during its internal audit
- As second-party audit
- By a customer to assess a supplier's operations

- As third-party audit
- By a quality system registrar as an input tool to the audit check-list

Who uses the QSA?

The QSA is a tool for any quality systems auditor.

Production Parts Approval Process (PPAP) Manual

What is the PPAP manual?

The PPAP manual describes the submission process of production and service commodities, including bulk material parts for approval to Chrysler, Ford, or General Motors. Its purpose is to ensure that all customer requirements are properly understood by the supplier and the supplier is capable of meeting these requirements. A software tool, the Form Completion Disk, is available at the AIAG to assist with the PPAP process.

Note: The PPAP process is a requirement by QS-9000 Section II, 1.1, which states that "Suppliers shall comply with all requirements set forth in the Production Parts Approval Process (PPAP) procedure."

When to use the PPAP

The PPAP procedure must be used in all situations occurring during the sample submission process.

Who should use the PPAP?

The PPAP process is usually the responsibility of the sales department, and the process is carried out in conjunction with the materials and engineering group.

In conclusion, it becomes apparent that all these manuals tie in with QS-9000 to form a complete quality system matrix (Figure 2.2).

Note: It is highly recommended that these reference manuals be treated as requirements. This is the preferred approach to using QS-9000.

2.3.2 How Should the Reference Manuals Be Used?

QS-9000 defines what is required. The reference manuals provide guidance on the specific approaches to fulfilling these requirements. The reference manuals are extremely useful documents and enable the user to develop a better quality system. Table 2.4 lists the QS-9000 requirements that each of the reference manuals can assist with.

Table 2.4: Reference Manual Cross-Reference

Reference Manual	QS-9000 Requirements
Advanced Product Quality Planning (APQP) and Control Plan	4.1, 4.2, 4.9
Potential Failure Mode and Effect Analysis (FMEA)	4.2, 4.4, 4.9
Measurement System Analysis (MSA)	4.11
Statistical Process Control (SPC)	4.9, 4.20
Quality System Assessment (QSA)	4.17
Production Parts Approval Process (PPAP)	4.2, 4.3, 4.4, 4.9, 4.13

For example:

Planning Section 4.2.3 states that "The supplier shall utilize the APQP and Control Plan reference manual."

Section 4.2.3 also states that for FMEA review and approval requirements, "Refer to the Potential Failure Mode and Effect Analysis reference manual."

Section 4.11.4 states, "The analytical methods and acceptance criteria used should conform to those in the Measurement Systems Analysis reference manual."

Section 4.20.2 states, "Consult the Fundamental Statistical Process Control manual."

Section 1.1 states, that "Suppliers shall fully comply with all requirements set forth in the Production Parts Approval Process manual."

2.3.3 Why Were the Reference Manuals Developed?

In the past, Chrysler, Ford, and General Motors each had their own expectations for supplier quality systems. There was little common approach. Differences between their expectations resulted in additional demands on supplier resources. In an effort to simplify supplier quality requirements, the purchasing vice-presidents of Chrysler, Ford, and General Motors chartered a task force to harmonize supplier quality system requirements including reporting formats and technical nomenclature through the development of these manuals.

2.4 ISO Guidance Standards

2.4.1 What Are the ISO Guidance Standards?

TC/176 has developed an array of guidance standards to supplement the compliance standards.

All standards (except 9001, 9002, 9003) are called guidance standards. Some aid in the selection of the appropriate compliance standard. Others provide the proper guidance for words and terms used in the quality manual, procedures, and work instructions. They expand the meaning and direction of the systems standards with in-depth information.

Core standards specify **what** must be done. Guidance standards provide insight into **how** to do it. Core standards are also very generic and not easily understandable. Several of the guidance standards have been developed to help with this issue.

2.4.2 What Guidance Standards Are Available?

ISO 8402 Quality Management and Quality Assurance—Vocabulary

Many ordinary words in everyday use are used in the quality field in a specific or restricted manner. The intent of this standard is to clarify and standardize the quality terms as they apply to the field of quality management.

Quality Management and Quality Assurance Standards

ISO 9000-1 Guidelines for Selection and Use

This standard provides guidance for the selection and use of the ISO 9000 family of International Standards on quality management and quality assurance.

ISO 9000-2 Generic Guidelines for the Application of ISO 9001, ISO 9002, and ISO 9003

The purpose of this standard is to enable users to have improved consistency, precision, clarity, and understanding when applying the requirements of quality system standards ISO 9001, 9002, and 9003.

ISO 9000-3 Guidelines for the Application of ISO 9001 to the Development, Supply, and Maintenance of Software

The process of development and maintenance of software is different than for most other types of industrial products. This standard provides the necessary additional guidance for quality systems where software products are involved.

ISO 9000-4 Guide to Dependability Program Management

This standard provides guidance on dependability program management and covers the essential features of such programs. In management terms, it is concerned with what has to be done, why, when, and how.

Quality Management and Quality System Elements

ISO 9004-1 Guidelines

This standard provides guidance on quality management and quality system elements. It explains each element of the quality system in detail, but the extent to which these elements are adopted depends upon factors such as market being served and nature of product.

ISO 9004-2 Guidelines for Services

This part of ISO 9004 gives guidance for establishing and implementing a quality system within an organization. It is based upon the generic principles of internal quality management described in part 1, and provides a comprehensive overview of quality systems specifically for services. The concepts, principles, and quality systems elements described are applicable to all forms of service.

ISO 9004-3 Guidelines for Processed Materials

This standard is a basic guide to quality management for processed materials.

ISO 9004-4 Guidelines for Quality Improvement

This standard is a guideline for management to implement continuous quality improvement within an organization. It explains in detail the actions to be taken throughout the organization to increase the effectiveness and efficiency of activities and processes to provide added benefits to both the organization and its customers.

ISO 9004-5 Guidelines for Quality Plans

This standard provides guidance to assist suppliers and customers in the preparation, review, acceptance, and revision of quality plans.

ISO 9004-7 Guidelines for Configuration Management

This standard provides guidance on the use of configuration management in industry and its interface with other management systems and procedures. It explains the management discipline required over the life cycle of a product to provide visibility and control of its function and physical characteristics.

Guidelines for Auditing Quality Systems

ISO 10011-1 Auditing

ISO 10011-1 emphasizes the importance of quality audit as a key management tool for achieving the objectives set out in an organization's policy. It establishes the basic audit principles, criteria, and practices and provides guidelines for establishing, planning, carrying out, and documenting audits of quality systems. It is sufficiently general in nature to permit it to be applicable or adaptable to different kinds of industries and organizations.

ISO 10011-2 Qualification Criteria for Quality Systems Auditors

ISO 10011-2 provides guidance on qualification criteria for auditors of quality systems. It is applicable in the selection of auditors to perform quality systems audits called for in ISO 10011-1.

ISO 10011-3 Management of Audit Programs

ISO 10011-3 provides guidelines for management of quality system audit programs defined in 10011-1.

Quality Assurance Requirements for Measuring Equipment

ISO 10012-1 Meteorological Confirmation System for Measuring Equipment

This standard contains quality assurance requirements for a supplier to ensure that measurements are made with the intended accuracy. It also contains an explanation for the implementation of these requirements.

Guidelines for Developing Quality Manuals

ISO 10013 Guidelines for Developing Quality Manuals

This standard explains the process for the development, preparation, and control of quality manuals. It presents the documentation hierarchy and formats typically used when documenting quality systems.

Note: In addition to the above quality system standards, ISO has published a position paper entitled Vision 2000, which explains the strategy for international standards implementation in the quality arena during the 1990s.

2.4.3 How Are the Guidance Standards Used?

The guidance standards are extremely useful documents and enable the user to develop a better quality system. The conformance standards state what is required, and the guidance standards assist you with the hows.

The recommended sequence for using the standards is as follows:

1. Use guidance standard 9000-1 to select the appropriate core standard.
2. Study applicable core standard 9001, 9002, and ISO 8402.
3. Read guidance standard (9004-1) to provide quality management guidelines.
4. Read ISO 10013 to develop quality system documentation.
5. Read ISO 10011 to develop internal audit program.
6. Read other ISO standards as necessary.

3 The Registration Process

3.1 Registrars

3.1.1 What Is a Registrar?

A registrar, also referred to as a "third party audit group," is an organization that is in the business of evaluating quality systems for compliance to the QS and ISO 9000 Standards.

3.1.2 What Does a Registrar Do?

The registrar conducts third party audits to assure that companies have quality systems that meet the QS/ISO 9000 quality system standards.

3.1.3 How is a Registrar Evaluated and Accredited?

For QS-9000, quality system registrars must be evaluated and accredited by a national body recognized by Chrysler, Ford, and General Motors (e.g., RvC, NACCB, RAB). For ISO 9000, quality system registrars are evaluated and accredited in most countries by an accreditation body established by the national authorities. Figure 2.7 shows the relationships of accreditation bodies to registrars.

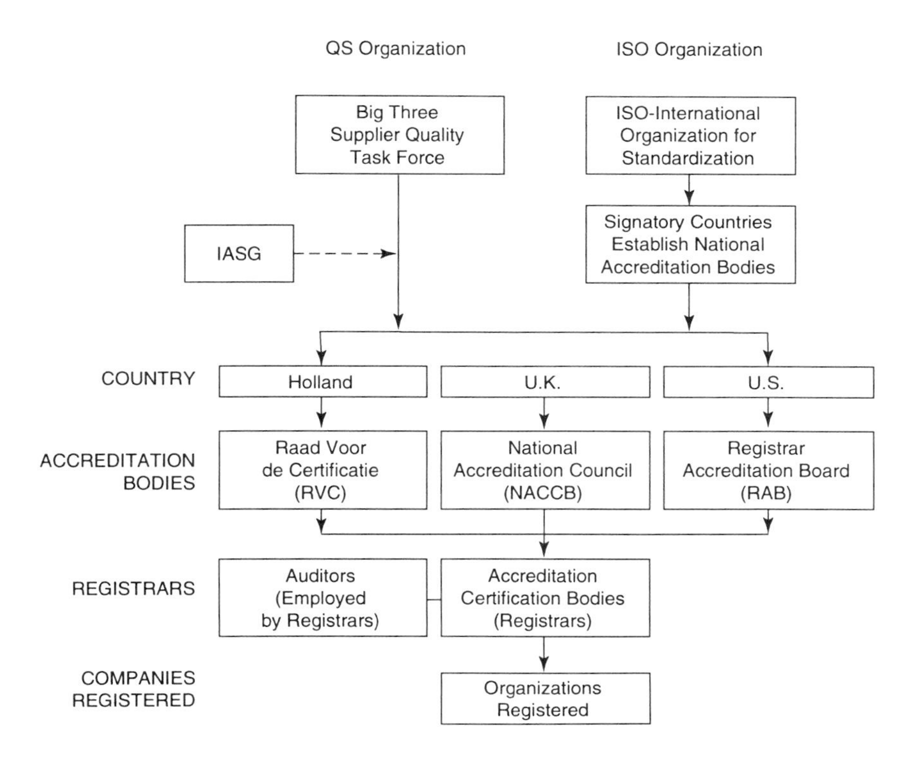

Figure 2.7 Relationships of Accreditation Bodies to Registrars

The standard used, EN 45012, defines the general criteria for certification bodies (ISO and QS) operating quality system certification.

Note: QS-9000 registrar requirements are defined in Appendixes A and G of the QS-9000 quality system requirements.

3.1.4 Who Are Some of the Registrars for QS-9000 and ISO 9000?

Table 2.5 lists the names of popular QS/ISO 9000 quality system registrars in the United States.

Table 2.5: List of Registrars

Name and Address	Tel and Fax
ABS Quality Evaluations, Inc. 16855 Northchase Drive Houston, TX 77060-6008	Tel: 713-873-9400 Fax: 713-874-9564
A.G.A Quality 8501 E. Pleasant Valley Road Cleveland, OH 44131	Tel: 216-524-4990 Fax: 216-642-3463
American Quality Registrars 1200 Main Street, Suite M107 Columbia, SC 29201	Tel: 803-779-8150 Fax: 803-779-8109
Bureau Veritas Quality International Inc. 509 North Main Street Jamestown, NY 14701	Tel: 716-484-9002 Fax: 716-484-9003
Det Norske Veritas Certification, Inc. 16340 Park Ten Place, Suite 100 Houston, TX 77084	Tel: 713-579-9003 Fax: 713-579-1360
Entela, Inc., Q.S.R.D. 3033 Madison, S.E. Grand Rapids, MI 49548-1289	Tel: 616-247-0515 Fax: 616-248-9690
Intertek Services Corporation 313 Speed Street, Suite 200 Natick, MA 01760	Tel: 508-647-5147 Fax: 508-647-6714

Table 2.5: List of Registrars (Continued)

Name and Address	Tel and Fax
KPMG Quality Registrar 150 West Jefferson, Suite 1200 Detroit, MI 48226-4497	Tel: 313-983-0344 Fax: 313-983-0500
Loydd's Register Quality Assurance Ltd. 33-41 Newark Street Hoboken, NJ 07030	Tel: 201-963-1111 Fax: 201-963-3299
OMNEX-Automotive Quality Systems P.O. Box 15019 Ann Arbor, MI 48106	Tel: 313-480-9940 Fax: 313-480-9941
Quality Management Institute Sussex Centre 90 Burnhamthrope Road West, Suite 300 Mississauga, Ontario L5B 3C3 Canada	Tel: 905-272-3920 Fax: 905-272-8503
Quality Systems Registrars, Inc. 13873 Park Center Road, Suite 217 Herndon, VA 22071-32279	Tel: 703-478-0241 Fax: 703-478-0645
SGS International Certification Services Meadows Office Complex 301 Route 17 North Rutherford, NJ 07070	Tel: 201-935-1500 Fax: 201-935-4555
S R I Quality Systems Registrars 2000 Corporate Drive, Suite 330 Wexford, PA 15090	Tel: 412-934-9000 Fax: 412-935-6825
TUV America, Inc. 5 Cherry Hill Drive Danvers, MA 01923	Tel: 508-777-7999 Fax: 508-762-8414
TUV Essen Olga Rada 2099 Gateway Place, Suite 200 San Jose, CA 95110	Tel: 408-441-7888 Fax: 408-441-7111
Underwriters Laboratories, Inc. 1285 Walt Whitman Road Melville, NY 11747	Tel: 516-271-6200 Fax: 516-271-6223

3.2 Registration Process

3.2.1 What Is QS-9000 and ISO 9000 Registration?

When an assessment of your company's quality system is conducted by a registrar and is found to be in compliance with the QS-9000 or the applicable ISO 9000 conformance standard, the registrar recommends registration to the accreditation agency.

The ISO 9000 and QS-9000 registrations can be concurrent but must be sequential.

3.2.2 What Are the Steps in the Registration Process?

Typical steps in the registration process are shown in Figure 2.8.

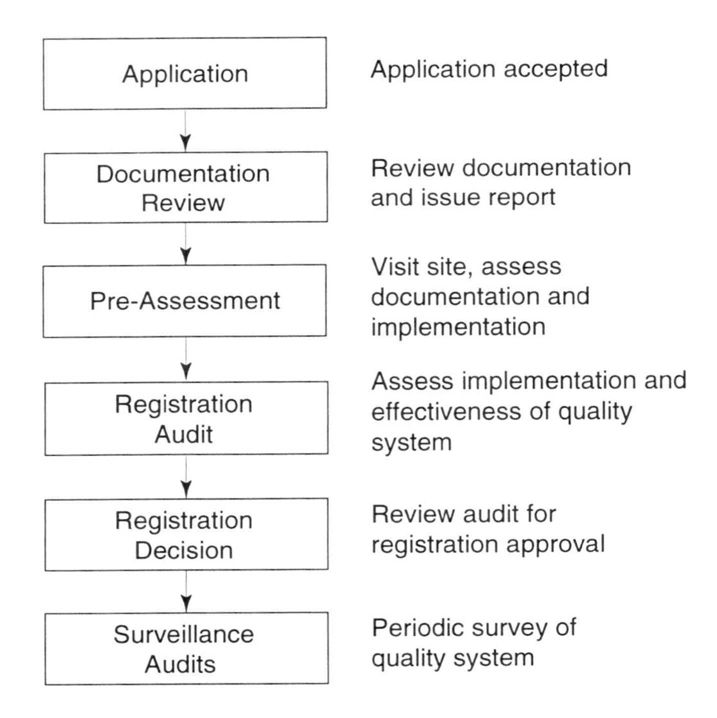

Figure 2.8 Registration Process

Application

Select the applicable QS-9000 requirements/ISO 9000 standard and the registrar to be used. Complete an application, providing basic information [i.e., company name, location, applicable standard, standard industry code (SIC), and statement of scope of registration]. Also include the number of employees and facility square footage affected by the scope of registration.

Quality System Documentation Review/Desk Audit

Supplier provides requested documentation to the registrar. The registrar reviews the documentation for compliance to QS-9000 requirements. A written formal documentation review report containing any inadequacies found are provided to the supplier. Normally the on-site audit will be delayed pending satisfactory resolution of the inadequacies.

Note: Some registrars prefer to perform this review at your facility. Others perform the review at their offices, saving travel costs and expenses.

Pre-Assessment (Mock Audit)

Pre-assessment or mock audit is a complete audit, determining if the quality system is ready for registrar examination. This audit can be performed by the internal audit team, a consulting team, or the registrar. Note: This is usually an optional activity.

Registration Audit

This phase determines the degree and effectiveness of the implementation of the quality system at the supplier's manufacturing and support locations. Based on the results of this audit, the registrar makes one of three recommendations.

- **Recommend Registration**

  This condition applies when no "major or minor non-conformances" were found during the audit.

- **Withhold Registration Pending Corrective Action**

  This condition applies when the applicant has open non-conformances which, in the judgment of the audit team, can be corrected by the applicant and be adequately verified by the audit team without a full re-audit.

- **Withhold Registration**

  This condition occurs when the applicant has one or more non-conformances which, in the judgment of the audit team, require a complete audit after corrective action.

Registration Decision

The registration board typically meets monthly and evaluates all audits and their respective recommendations. Upon satisfactory completion, registration is granted and a certificate of registration is issued.

Surveillance Audit

Surveillance audits must be performed every six months by the registrar to check for continued compliance.

Note: See Chapter 5 for details on the various audits.

3.2.3 What Are the Causes of Failed Registration Efforts?

Registration efforts fail for any number of reasons. The top four reasons are:

- Lack of management support
- Lack of training
- Poor quality system documentation
- Inadequate corrective actions

3.2.4 How Long Does It Take to Get Registered?

The state of the existing quality system is the primary factor in determining the time required for registration. Other factors, such as the company's commitment and the resources it is willing to expend, also play an important role.

A well-documented system may require minor changes and the registration effort may take as little as six months. But, on the average, registration time is slightly longer than a year.

3.2.5 How Much Will a Registration Effort Cost?

Costs vary widely, although the internal costs will far exceed the costs of the third party activity. To those who challenge the internal costs, it is a popular view that these activities are quality related and something that the organization should have been doing all along.

External costs vary; they depend upon the size and type of facility, and normally range from $10,000 to $30,000 per facility.

4 IASG-Sanctioned Interpretations

The following excerpts are from the IASG Sanctioned Interpretations dated 11/9/95 and are reprinted by permission of the IASG and The Supplier Quality Task Force. The date of issue of these interpretations is important, as they may be periodically revised or updated. All references are to QS-9000, 2nd ed. 1995, unless otherwise stated. (The questions are in bold faced type, with the answers or actions in normal faced type.)

What are the IASG Sanctioned Interpretations?

IASG Sanctioned QS-9000 Interpretations are QS-9000 related questions that are sanctioned and recognized by the Chrysler, Ford, General Motors Supplier Quality Requirements Task Force, the participating ISO 9000 accreditation bodies and QS-9000 qualified registrars.

Where can I get a copy of the latest IASG Sanctioned Interpretations?

Copies of the latest IASG Sanctioned Interpretations are available through the American Society for Quality Control (ASQC). ASQC can be contacted at 1-800-248-1946.

The interpretations are also published by several magazines, including The Quality Systems Update.

How are the IASG Sanctioned Interpretations prepared?

1 All IASG QS-9000 interpretations must be processed at the issue level as follows:

   a New Issue presented to the IASG for discussion - May include only the question.

   b Draft language distributed to the IASG members for consensus - This would include questions and draft answers by members of the IASG or from a submission.

   c Agreed status is achieved after consensus of all members - the Agreed date applied is the meeting date.

   d Incorporation into the IASG Sanctioned QS-9000 Interpretations document.

   e The sanctioned interpretations document is distributed to stakeholders, IASG members, all QS-9000 recognized accreditation bodies, all accredited registrars associations with membership represented and the public.

2 Representatives from Chrysler, Ford and GM must, individually, agree with interpretations and IASG decisions prior to completing Step #c above.

3 All discussions, tentative decisions, and minutes resulting at and from the IASG meetings are considered confidential to the working group, and are treated as such until the Agreed status is reached and Step #e above is initiated.

4 The IASG retains final approval of IASG membership, configuration and size of the group. No substitutes, alternates or back-up company representatives are permitted to attend.

5 Regular attendance at IASG meetings is critical and is expected. Repeated absences may result in being replaced as a working member of the IASG. The IASG will typically schedule at least three meetings in advance of a current meeting.

How are the IASG QS-9000 interpretations classified?

They are grouped by categories:

A= Applicability

B= Appendix B: Code of Conduct

C= Criteria: Subdivided by the 24 QS-9000 Elements within Sections I, II, III

D= Database

O= Other

P= Process

R= Registration/Accreditation

T= Training

APPLICABILITY (A)

Can anyone be certified by a QS-9000 qualified third party registrar to QS-9000?

Only those suppliers meeting the "Applicability" definition are required to achieve compliance/registration. Any supplier or subcontractor may elect to pursue third party registration; however, to obtain QS-9000 registration, all elements of QS-9000 must be assessed and complied with. Only "Servicing" (and Design Control for ISO 9002) may be determined as not applicable by the registrar. The registrar must assure the Code of Practice, Appendix B, requirements are met.

Can a QS-9000 (ISO 9000) certificate be issued by anyone other than a "QS-9000 qualified" third party registrar?

No. The term QS-9000 is protected as property of the Big Three. Only those third party registrars meeting the QS-9000 qualification requirements of at least one Big Three recognized accreditation body are permitted to issue a registration certificate with the term QS-9000.

Who are the Big Three recognized accreditation bodies for QS-9000?

Currently, JAS-ANZ, UKAS, RAB, RvA, SWEDAC, TGA, SAS, SINCERT, ENAC, FINAS and SCC. Additional accreditation bodies worldwide will be added in the future.

What is the status of the European launch of QS-9000?

It is well underway. QS-9000 qualified accreditation bodies are established, and many registrars have either applied for, or achieved, QS-9000 qualification. European representatives, including registrars and accreditation bodies, now participate on the IASG. Registrar auditor and supplier QS-9000 training courses are being provided.

QS-9000, August 1994 states that registrars be permitted to use a full QS-9000 or an ISO 9000 upgrade to QS-9000 as a witness assessment.

QS-9000, Appendix B, page 81, states that suppliers registered to ISO 9000 without consideration of QS-9000 ...shall update the quality system documentation... as a part of being registered to ISO 9000 and QS-9000. This is clearly an upgrade option for currently ISO 9000 registered companies.

However, the above referenced sentence goes on to say that suppliers choosing registration shall... identify these revisions to the registrar at the next surveillance visit. This appears to undermine or usurp the individual customer's process of implementing QS-9000 third party registration by selecting a date by which suppliers must be registered as opposed to at the next surveillance.

While the Chrysler/Ford/General Motors expectation is full compliance to QS-9000, a supplier is permitted to use a two step process involving ISO 9000 registration followed by a QS-9000 upgrade.

We are a Tier 1 supplier to the automotive OEMs. How do we determine if we have design responsibility?

If the supplier has the authority to establish a new product specification, or change an existing product specification, for any product a supplier ships to an OEM customer, then they are design-responsible. The requirement for customer approval of changes does not affect this. Consult with your customer engineering activity for further clarification.

Does the semiconductor supplement from Delco Electronics, Ford and Chrysler have to be used for a QS-9000 audit of a semiconductor supplier to the Big Three?

Yes, the supplement says that QS-9000 and the supplement becomes the requirements for semiconductor suppliers. NOTE: This does not extend to other electronic products beyond semiconductors.

Could you define the difference between product and material as used in doing Quality System Assessments. We are jointly owned by two companies; we receive steel coils from these two companies, plate zinc onto the strip surface, rewind coils, and ship to OEM blanking or stamping facilities as directed by our owners.

You appear to be a toll processor, i.e., a subcontractor, to Tier 1 supplier(s). The product and material may appear to be one and the same. In general, product is what you sell, material is, in part, what you use to produce the product you sell.

APPENDIX B: CODE OF CONDUCT (B)

A supplier used a registrar's sister subsidiary as a consultant prior to ISO 9002 registration. Subsequently, that registrar registered the client. Can the same registrar be used to upgrade the client to QS-9000?

Yes. If consulting actions predate the August 1994 release date of QS-9000. All consulting must have ceased by August 1994.

We plan to use (X) registrar as our QS-9000 registrar, however, (X) is not on the current list of accredited registrars. Will an (X) certification assessment be sufficient for us as a Tier 1 supplier to the Big Three? Should we be concerned that they are not on the list?

They must be on the list; if not, they are not permitted to issue a QS-9000 registration certificate. However, registrars having achieved Step A.2 of Appendix G, QS-9000:1995, Second Edition, are permitted to complete no more than four assessments for QS-9000, one of which must be witnessed by the accrediting body.

CRITERIA (C)

Section I. ISO 9000-Based Requirements

4.1. Management Responsibility

Must all elements of the QS-9000 standard be included in the management review process (4.1.3)?

Yes, even if the element does not specifically state that. For example, in 4.14.3 Preventive Action, it states that plans and results must fall under management review... BUT Section 4.14.2 Corrective Action does not state it specifically. Despite this inconsistency in the ISO 9000 standard, corrective action plans and results must also be part of management review.

The accreditation bodies, registrars, and OEM's all agree that the intent of the standard is to require all relevant ISO 9000 systems be included in management review; likewise all QS-9000 systems must be included in the management review process.

Should the business plan be a controlled document with rev#, date, distribution list, etc.?

Yes, business plans have to be controlled documents.

In the recently issued interpretations dated August 1, 1995, it was stated that business plans MUST be treated as Controlled Documents. I would like to know if an alternative approach of treating these documents as Quality Records would also be acceptable?

The ISO 9000 definition of records does not include business plans; QS-9000 does require that Business Plans be controlled documents.

4.2 Quality System

Does QS-9000 require that flow charts be included in the policy and/or procedures manuals?

No, but they are welcomed by most auditors because they tend to clearly identify who, how, the interrelationship, and the sequence. However, flowcharts are a PPAP requirement.

When documenting our Quality Policy Manual for QS-9000, is it required to include a responsible department/individual?

Yes. See ISO 10013 Guidelines for Developing Quality Manuals, and ask your registrar. Also refer to QS-9000, cl. 4.2.1 and the QS-9000 glossary under quality manual.

4.4 Design Control

Reference Sub-elements, Section I.4.4.4, requirements for computer-aided design and analysis and the I.4.15.6 requirements for computerized system for ASN's. We believe these are particularly uncommon and onerous for many of the subcontractors. How are these requirements to be interpreted by registrars?

Sub-elements 4.4.4 and 4.15.6 can be waived by the customer per QS-9000. Objective evidence of a waiver, if applicable, must be available to show the auditor.

4.6 Purchasing

Regarding 4.6.2 (QS-9000:August 1994) Scheduling Subcontractors - suppliers shall require 100% on-time delivery performance from subcontractors... This is not consistent with 4.15.6 which requires a GOAL of 100% on-time delivery. Since delivery is a performance factor, should not the requirement of subcontractors be "a goal of, and a delivery system with the capability and design to be able to, provide 100% on-time delivery"?

Yes. Following is the new QS-9000:Feb 1995 text:

4.6.2 Scheduling Subcontractors

Suppliers shall require 100% on-time delivery performance from subcontractors. The supplier shall provide appropriate planning information and purchase commitments to enable subcontractors to meet this expectation.

The supplier shall implement a system to monitor the delivery performance of subcontractors, including tracking of premium or excessive freight.

4.15.6 Supplier Delivery Performance Monitoring

(Paragraph 1) The supplier shall establish systems to support 100% on-time shipments to meet customer production and service requirements. If 100% on-time shipments are not maintained, the supplier shall implement corrective action to improve delivery performance, including communication of delivery problem information to the customer.

(Paragraph 2) A supplier shall have a systematic approach to develop, evaluate and monitor adherence to established lead time requirements. The supplier shall implement a system to track performance to the customer delivery requirements.

What is meant by QS-9000 Clause 4.6.2, "Suppliers shall perform subcontractor development using Section 1 and 2 of QS-9000"?

A definition of OEM customer-approved second party will be handled on a case-by-case basis.

The new wording in Section 4.6.2 of QS-9000:Feb 1995 now reads as follows:

Subcontractor Development

Suppliers shall perform subcontractor quality system development using Sections I and II of QS-9000 as the fundamental quality system requirement. Assessments, if part of subcontractor development, should occur at supplier specified frequency. Subcontractor assessments to QS-9000 by the OEM customer, an OEM customer-approved second party, or an accredited third party registrar (see Appendix B) will be recognized in lieu of audits by the supplier.

Also, of note is the definition of Subcontractors in the Second Edition of QS-9000: Feb 1995:

Subcontractors

Subcontractors are defined as providers of production materials, or production or service parts, directly to a supplier to Chrysler, Ford or General Motors or other customers subscribing to this document. Also included are providers of heat treating, painting, plating or other finishing services.

The suppler must communicate QS-9000 as defining the supplier's fundamental quality system requirements to all subcontractors. Additional initiatives that show subcontractor development could include, but are not limited to: a) training on QS-9000, b) remedial actions taken on non-conformities using QS-9000, c) assessment of subcontractors using QSA (Appendix A), d) requiring of QS-9000 registration.

All elements of QS-9000 should be applied to subcontractors recognizing that suppliers may supplement QS-9000 with definitions or methods which are company-, division-, or commodity-specific to their subcontractors.

What constitutes subcontractor development? (4.6.2) Does the level of development depend on the importance of the subcontractor in processing the final product?

Subcontractor development, as defined in QS-9000, refers to all activities designed to improve the fundamental quality system performance of the subcontractor (as defined in the QS-9000 glossary). The level of development is dependent upon the needs of the subcontractor relative to the requirements of QS-9000 and the importance of the product or process they supply. Deployment of QS-9000 through contracts, workshops, surveys, corrective/ preventive actions and documentation requirements are all considered acceptable forms of subcontractor development.

4.9 Process Control

Define environment (as used in 4.9(b), suitable working environment).

Environment will vary for each site, but generally includes: housekeeping, lighting, noise, HVAC, ESD controls, safety hazards relating to housekeeping. Environment is defined in the Glossary.

Clause 4.9.6 PPAP, page 2, Items 7 and 9. What constitutes any change in process? What constitutes change in subcontracted parts, materials or services?

The process has changed when any of the following are changed: part number, engineering change level (from the design record), manufacturing location, material/component subcontractor, or production process environment. If in doubt, notify the customer for their position. See definitions of Process and Subcontractor in PPAP manual.

4.12 Inspection and Test Status

Clause 4.12 under Product Location states: "location of a product in the normal production flow does not constitute suitable indication of...status unless inherently obvious...."

Considering current production and inventory methods of KAN BAN, bar codes, cellular manufacture, etc., can this clause be strictly enforced to require additional tags, etc., on baskets, totes, product?

Latitude is permitted, beyond automated production transfer processes, if the test status is clearly identified, documented, and achieves the purpose (i.e., known status).

4.13 Control of Nonconforming Product

Regarding 4.13.4, wherein prior written customer authorization is required whenever the product or process is permanently changed from that currently approved... Does a verbal phone authorization from the customer, documented by the supplier, constitute "written" authorization?

It is acknowledged that this practice was used by many OEM's. The answer, however, is no. Temporary changes to process may be verbally authorized with follow-up documentation. All permanent changes must have prior written authorization.

4.16 Control of Quality Records

The second revision attempted to better clarify record retention periods, however, the addition of the word minimum is confusing, e.g., can a record be kept indefinitely?

Retention periods longer than those specified in QS-9000, Clause 4.16 can be specified by a supplier in their procedures, but records must eventually be disposed of in order to comply.

Control of Quality Records, Element 4.16, under Record Retention refers to..... purchase order and amendments shall be...... Can you please clarify whether the purchase order in this context refers to:

a) purchase orders and amendments placed on the supplier by the customer (i.e., the Big 3);

b) purchase orders and amendments placed upon a subcontractor by the supplier; or

c) both of the above.

Records retention reference to Purchase Orders and Amendments includes those issued both to and by the supplier, i.e., c) both of the above.

4.17 Internal Quality Audits

Element 4.17, Internal Quality Audits, Page 46, states:

Internal quality audits shall be scheduled on the basis of the status and importance of the activity to be audited...

Our question is: In this case, what does the use of the term activity refer to? Is activity, for example, referring to a product or manufacture area? Or is activity referring to a function within the QS-9000 plant quality system such as MSA, PPAP, FMEA, control plans, etc.?

We are interested in the IASG interpretation of activity as this has a direct effect on audit scheduling.

Activity can refer to both departments and processes in a company. The internal audits must include all processes and procedures implemented to address all elements of QS-9000. Internal audit results shall be included in Management Review (4.1.3) to be evaluated for continuing suitability and effectiveness.

4.18 Training

How can confirmation of training effectiveness be demonstrated as required by QS-9000 4.18? This question is directed at learning whether the automakers have a particular method in mind.

Training effectiveness can be best judged by the performance of the trained individuals via audits and performance evaluations. See your registrar. The automakers have no particular method in mind.

4.19 Servicing

The supplier's component is part of a sub-assembly at the customer which in turn is attached to the vehicle. It is not repairable by the dealer network, only replaceable. The supplier does provide engineering design support, warranty analysis and subassembly component interface investigation (review for non-conformances at the OEM). What supplier activities mentioned above, if any, are considered covered by QS-9000, Section 19, Servicing?

None. Any after-sales product servicing provided as part of the OEM contract or Purchase Order would fall under **Element 4.3**.

Section II. Chrysler, Ford and General Motors Requirements

2.1 Production Parts Approval Process

Relative to PPAP, a QS-9000 applicant has continuously supplied products to the OEMs since 1987, having met all sample submission requirements, and having no interruptions or changes. They have not completed any PPAPs, nor have they been requested to do so. Is there anything else they must do to comply with QS-9000 requirements?

If there have been no changes in "part number, engineering change level, manufacturing location, material subcontractors or production process environment" since 1987, then no PPAPs would be expected unless specifically requested/notified by the OEM customer. The system for implementing PPAP must be in place. The registrar will expect to see evidence of PPAP implementation for all parts submitted since September 1993. Pre-PPAP part submissions must show compliance to the then existing customer requirements.

Supplier has PPAP process documented adequately, and if he is requested to submit parts for approval, the documented process will meet the requirements. The supplier provides "off-the-shelf" items they design for customers. Their only OEM customer has issued PPAP approval documents showing part is approved without requiring the supplier to do the PPAP requirements.

Should the registrar accept this and recommend for registration to QS-9000?

Supplier must meet all required steps according to PPAP or the previous customer requirement in effect, even if request is waived. PPAP file must be available for registrar or customer review and show compliance to part submission requirements in effect at time of submission.

Must a company obtaining QS-9000 certification, require Section II.1, PPAP, of their subcontractors?

It varies. All GM suppliers as defined in QS-9000, who are required to be certified to QS-9000, must require PPAP of their commodity subcontractors, as indicated in General Motors Operating Policy for PPAP on page 32, Appendix D, of the PPAP manual. For bulk, raw, or indirect material, it is the Procuring Division's decision whether PPAP is required.

For all other organizations who are QS-9000 certified, the PPAP requirement for their subcontractors, must be treated by them as a shall or a should depending on their OEM customer requirements or directives. If not specifically required by the OEM customer, then it is a should, where PPAP is a preferred subcontractor methodology, that can be replaced by an equal but more appropriate approach.

2.2 Continuous Improvement

QS-9000 Section II, 2.1, first paragraph includes price as a continuous improvement factor. Registrars need guidelines of whether they can or should audit this and, if so, what criteria? This seems to equate price reduction with continuous improvement.

Auditing of specific part price information is not expected for third party quality system assessments. However, the use by a supplier of cost elements or price as one of the key indicators within a continuous improvement system is required and subject to registrar audit.

2.3 Manufacturing Capabilities

Ford and Chrysler do not require lab accreditation (QS-9000 4.10.1 sub). Does GM still require lab accreditation if a supplier passes a third party assessment to QS-9000 by a "Qualified QS-9000 Registrar"?

GM has changed its policy for laboratory accreditation. Third party registration to QS-9000 in accordance with Appendix B will satisfy the GP-10 requirements for GM North American locations of laboratory facilities utilized by suppliers for inspection and testing of their own product for purposes of conformance to the specified requirements. Laboratories utilized for commercial laboratory services are excluded from this provision. This practice, as stated, is not acceptable and does not meet QS-9000 4.9.

DATABASE (D)

How are suppliers expected to stay informed of changes to documents, i.e., PPAP? Is a subscription service available?

No subscription service is available at this time; please contact your customer's purchasing activity to verify the latest dates.

A) How can I get a list of the QS-9000 registered companies? B) Who collects this information? How is it collected? C) How many companies are currently registered to QS-9000?

A) You can't at this time. B) The IASG collects the names of QS-9000 registered companies through submissions by QS-9000 qualified registration/certification bodies to the Fax Mailbox (USA): 614-847-8556. The list is maintained at present by each of the Big Three IASG representatives. Access by the public is planned through publicly accessible database(s) in the near future. C) As of 11/9/95, the number of QS-9000 certificates issued number less than 60.

OTHER (O)

We heard of a "TE-9000" for automotive suppliers of tooling and equipment...what is it?

A modified version of QS-9000 is being developed for auto suppliers of tooling and equipment. TE-9000 is a draft name of a QS-9000 based quality system requirements document. The term TE-9000 is an internal Chrysler/Ford/General Motors Supplier Quality Requirements Task Force (copyrighted) term at this time.

More information will be available in the AIAG Actionline soon. Third party registration to TE-9000 may or may not be required.

Should suppliers of fixtures and gauges wait for TE-9000? Are they required to implement QS-9000 if they supply fixtures and gauges directly to the Big Three?

No company should wait for TE-9000. If it makes sense, supplier management should consider pursuit of ISO 9000, (not necessarily registration), if their company is not otherwise eligible for QS-9000.

PROCESS (P)

Will the Big Three accept any ISO 9000 or QS-9000 certificates based primarily on first party internal audits, as described by Hewlett-Packard/Motorola representatives, in their Supplier Audit Confirmation (SAC) Approach?

No. The Chrysler, Ford and GM Supplier Quality Requirements Task Force have issued a position paper to registrars and accreditation bodies to document the Big Three positions relative to the proposed Supplier Audit Confirmation (SAC) approach presented at the 1/19/94 IAF Conference in Geneva. In essence, the Big Three do not accept any first party declarations of conformance to QS-9000. Nor do they accept a third party assessment which does not meet the QS-9000 requirements; the latter includes the assessment of all quality system elements by a QS-9000 qualified assessor working for a QS-9000 qualified registrar.

Must auditors always report "Opportunities for Improvement" for a QS-9000 assessment?

From QS-9000 App. B, Item 8. "..Third party auditors will identify opportunities for improvement (e.g., excessive scrap) as these become evident during the audit without recommending specific solutions. These opportunities shall be included in the report to the supplier."

Opportunities for improvement (see Continuous Improvement 2.2 of the standard) are expected of the auditor. If none are found, a statement to that effect must be reported.

How should a QS-9000 auditor address customer performance requirements?

In QS-9000, supplier internal key indicators must be established to meet customer performance requirements, e.g., Analysis and Use of Company-Level Data (4.1.5), Customer Satisfaction (4.1.6), On-Time Delivery (4.15.6) and Continuous Improvement (2.1). Effectiveness of a company's system must be measured and tracked by the use of these key indicators. Continued surveillance by an auditor of poor trends in those key indicators in terms of meeting customer performance requirements will jeopardize continued QS-9000 certification.

We are going through a QS-9000 audit in mid-September. When will Section III requirements at the audit be imposed? Will Section III requirements be audited as sampled?

All sections will be audited during the registration audit. Conformance to Section III requirements will be evaluated under element 4.3 (Contract Review). The registrar must ascertain which of the Section III requirements are applicable to you based on your automotive customers; this should occur at the pre-audit visit, or sometime before the registration audit. Each applicable item in

Section III must be audited during the initial audit or in the surveillance visits over the subsequent three-year period (see QS-9000 Appendix B, Item 7). The auditors should include the principal Section III items at the registration audit.

Is there an approved checklist of questions available for QS-9000 auditing?

Yes. QSA (Appendix A, QS-9000:1995) is an approved checklist, however, it is not comprehensive nor is it intended to completely prepare a supplier for QS-9000. Suppliers and registrars should supplement the QSA with additional auditing material to assure conformance with all elements of QS-9000.

How can the IASG help reduce auditor inconsistency relative to inspection and testing, e.g., in-house lab facilities?

Auditor teams for QS-9000-qualified registrars must be qualified to audit in-house lab facilities in order to audit compliance to QS-9000, including clauses 4.10 and 4.11. Auditor on-site verification must include:

-- adequacy of the laboratory procedures

-- qualifications of the lab personnel conducting tests

-- conducting of the appropriate tests for the commodity(s)

-- performing these tests correctly, to the appropriate process standard, e.g., ASTM.

Accreditation bodies must provide competent auditors for the registrar witness audits and verify that adequate time is devoted to the audit of the in-house laboratories by registrars.

Reference QS-9000, Appendix B, Item 7: The entire quality system shall be assessed at a minimum of once every three years. It is permissible for each surveillance

audit to re-examine part of the system so that the equivalent of a total re-assessment is completed within each three year cycle. This could be interpreted in several ways:

1. **Every three years a re-approval assessment is performed with a reduced intermediary surveillance duration; or**

2. **Extended surveillance is performed in lieu of the three year re-approval.**

Does the matrix in Appendix H apply to the extended surveillance (2.) situation? If it does, then what should be the reduced surveillance duration where a re-approval assessment is to be performed every three years?

If there is no reduction in the surveillance duration where a three yearly re-approval is performed, then this is tantamount to making the second option mandatory, which appears contrary to the statement in Appendix B, Item 7.

Appendix H defines the MINIMUM audit-person days required for all initial and surveillance visits regardless of the registration cycle or surveillance approach.

Item 12 on Page 80 of QSA states Registrar's checklists <u>shall</u> include all questions contained in the QSA. Numerous QSA questions relate to should items in QS-9000. Page 1 of QS-9000 states, the word "shall" indicates mandatory requirements. The word "should" indicates a preferred approach.

By their inclusion in the QSA, are these "shoulds" now elevated to mandatory requirement status from a

preferred approach? If not, why are they included in the QSA while other shall requirements are omitted?

No, however, a "should" statement is a requirement with some flexibility allowed in compliance methodology. An alternate method of satisfying the intent of the "should" requirement can be acceptable...but the "should" must be satisfied.

See QS-9000 Appendix B, Item #12. They are included in the QSA because there was no intention for the QSA to be all inclusive of QS-9000 requirements. To provide full coverage of the QS-9000 requirements, first and third party auditors must supplement the QSA questions with their own questions.

How will the registrar audit Section III?

The registrar will utilize Contract Review, 4.3, to audit compliance to Section III. All Section III requirements that apply should be sampled at least once over the three year contract period by your registrar, starting with the registration audit. Pre-audit document review and/or any pre-assessment should address relevant Section III requirements.

How far are auditors allowed to delve into the business plan?

They must verify that the supplier is conducting strategic business planning, with appropriate initiatives as defined in the QS-9000 Business Plan requirement. Often a review of evidence such as dated Tables of Contents, and a review of a few non-sensitive sections is sufficient confirmation that policies and procedures are being followed.

The situation is as follows: 15 manufacturing sites are all planning to receive separate QS-9000 certificates, and these 15 manufacturing sites depend on a design

function at yet another site that supports ALL 15 manufacturing sites. The design site would like to keep its assessments to a practical number, and would therefore like to be independently registered to QS-9000. The concern is that QS-9000 wording (Appendix B, #2) sounds like a technicality that would require that the design function would need to be visited for registration and surveillance for all 15 sites registrations. Is this true?

QS-9000 requires that only those sites defined in the applicability section may be registered, therefore off-site locations, e.g., design, purchasing, will not receive QS-9000 certificates. The design center support for all 15 plants can be audited in the initial audit, and be put on a regular six-month surveillance plan, if the same registrar is contracted for all the sites. Design functions audited can be tracked by the registrar on their audit matrix, so additional design center audits would not be necessary.

REGISTRATION/ACCREDITATION (R)

If a multi-site corporation has a design engineering center (DEC) and three manufacturing sites (M1, M2, M3), and is "design responsible" to Chrysler/Ford/General Motors, at what point can the QS-9000-qualified registrar issue an ISO certificate with QS-9000 notation? (They are seeking individual site certificates.)

Design responsible suppliers cannot achieve QS-9000 registration at the ISO 9002 level.

Even if site M1 is the first assessed and registered to ISO 9002 and complies to all QS-9000 elements except 4.4 Design, it cannot be granted QS-9000 because the company is considered design-responsible. Even though an ISO 9002 certificate can be granted, the Big Three requires the DEC be audited and in compliance to

Section I. 4.4 Design, and all other applicable elements, before any reference to QS-9000 compliance can be added to <u>one or any</u> ISO 9001 certificate.

Item #11 of the Accreditation Bodies' QS-9000 Documentation Information Questionnaire, required of a QS-9000 registrar, states "each site (must) be individually registered and therefore individually audited, regardless of the type of audit; and acknowledges that sampling of sites is not allowed." Does this mean individual certificates? Does the sampling restriction apply to all types of company locations?

Individual certificates are not required, but every site must be assessed, and each site must appear listed on a certificate.

Insert the words <u>design/manufacturing</u> before the word "sites." Sales or distribution locations may be "sampled" according to the EAC Guidelines of May 1994.

What must the certificate with QS-9000 notation have?

The QS certificate must meet all requirements of a typical ISO 9000 certificate and, in addition:

a Must include all products and services being supplied to one or more of the companies subscribing to this document,

b Cite a separate QS-9000 scope, date of issue, product line(s), statement as to what was audited for QS-9000 - because the scope and duration of QS-9000 certification may be more limited than that of the ISO 9000 registration (Appendix B, Code of Practice, Item 8.),

c Include terms to appear somewhere on the first page: "having been audited in accordance with the requirements of QS-9000 Appendix B, Code of Practice

d Have as much of the above as possible on its face. The company name, standard, address, and dates of registration must appear on the front page. If an attached schedule is needed, it must be referenced or noted on the first page,

e Include every registered site and its location, and scope must be listed.

Will Big Three recognized accreditation bodies recognize each other's witnessing of QS-9000 assessments?

The Big Three encourages recognition arrangements. At this time, RAB and RvA will recognize one another's witnessing of QS-9000 step two audits. Each, however, requires step one (application and documentation) to be completed and approved independently, and each requires receipt of acceptable witnessing documentation and results regarding step two from the other.

Certificates for ISO 9001/2:1994 with a QS-9000 certification notation cannot be issued with the mark of a particular accreditation body until that accreditation body:

a is recognized by the Chrysler/Ford/GM Supplier Quality Requirements Task Force,

b has received and accepted step one information from the registrar,

c has either completed step two, or has received acceptable step two information from another recognized accreditation body, and

d has notified the registrar that "QS-9000" certificates can be issued.

Have any of the auto companies endorsed or supported any particular registrars for their suppliers?

No! All registrars successfully completing the agreed upon qualification steps will have their certifications recognized by the Big Three. The accreditation bodies (currently recognized) will provide the "QS-9000 qualified registrar" lists to the Big Three; no other sources can.

Question - The minimum mandate requirements for on-site auditing are given in a recent issue of EN 45012 EAC Guidelines. Are there man-day guidelines for QS-9000?

Yes. The QS-9000:February 1995 release includes a Survey Audit Days Table in Appendix H. It has been modified several times since originally included in our IASG release. Please review the changes made since our last 9/28/95 IASG release - they have been underlined.

Appendix H:

Survey Audit Days Table

Table 2.6 shows the MINIMUM number of man days which should be spent by the registrar on initial QS-9000/ISO 9001 quality system audits (see Glossary) and ongoing six-month surveillance audits (see Appendix B, Item 7). The MINIMUM number of man days for QS-9000/ISO 9002 audits may be reduced by 20%. Registrars will document actual on-site audit man days, including any deviation below the MINIMUM. Accreditation bodies will review such documentation for appropriateness.

Use of this table by registrars is effective immediately and remains in effect until modified by the Supplier Quality Requirements Task Force.

Table 2.6: Survey Audit Days

Certified Entity: Number of Employees	Initial Audit (on-site man-days)	Ongoing Six-Month Surveillance Audits (on-site man-days)*
1 - 15**	2	1
16 - 30**	4	1
31 - 60**	5	1.5
61 - 100**	6	1.5
101 - 250	8	2
251 - 500	10	2.5
501 - 1000	12	3
1001 - 2000	15	3.5
2001 - 4000	18	4.5
4001 - 8000	21	5.5

Source: Based on EAC Guidelines on EN 45012, Draft, May 10, 1994.

*Minimum man-days for the first three years after QS-9000 registration.

** Revised 8/25/95.

The above table was developed to primarily apply to one site/one certificate situations.

Notes on QS-9000:Feb 1995 Survey Audit Days Table:

1 Initial Audit (On-site Man-Days) <u>cannot</u> include pre-audit document review (whereas the EAC Guidelines do).

2 <u>Initial Audit (On-site Man-Days)</u> cannot include pre-assessments which are provided for supplier feedback only, with non-binding review, and corrective actions that are not part of the registration audit (don't appear in the final report).

3 <u>Initial Audit (On-site Man-Days)</u> a) can include single or multiple registration audit visits which occur less than three months after document review and the audit matrix are completed, b) do include binding nonconformances leading to,

c) approved corrective actions which are included in the final registration audit report, and d) utilize the same qualified QS-9000 auditors for each visit or step.

4 Audit man-days for registration upgrades from ISO 9001/2 to QS-9000 are not addressed.

5 It is expected that the audit man-days will include auditing on all shifts.

In summary, only those man-days subsequent to completion of the document review, and development of the audit matrix, and that occur within a consecutive three month period may be counted as man-days in accordance with the Appendix H Table (Figure 2.9).

The registrar should treat these man-days as true minimums. If the days quoted are below the minimums stated, the accreditation body shall assess the validity of such justification. (Refer to Accreditation Body Notification which follows.) The actual on-site initial audit man-days must be reported in the QS-9000/ISO 9001/2 registration report.

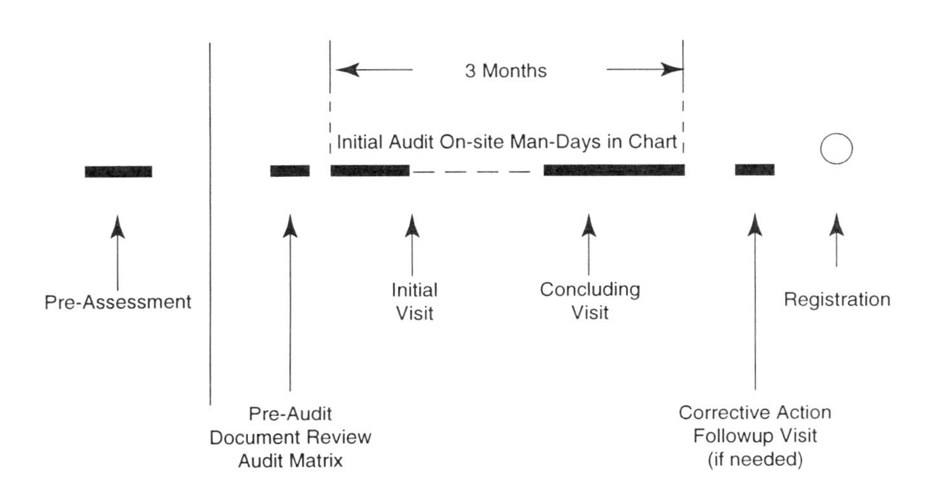

Figure 2.9 Audit Man-Day Timeline

Chart Definitions

Column #1 of Appendix H, entitled Certificated Entity: Number of Employees, represents the total number of employees per site including all shifts, and all administrative, professional, etc.

Column #2 of Appendix H, entitled Initial Audit (On-site man-days), represents the minimum number of audit man-days for a site undergoing a single certificate site audit. Time required for documentation review is in addition to these days.

Sites are defined as locations at which production processes occur; corporate schemes apply only to multiple site registrations. Remote locations, e.g., Engineering, Purchasing, must be audited as they support a site(s), but man-days to conduct these audits are included in a site audit as defined in the Appendix H Table.

Corporate/Multi-Site Considerations

In multi-site situations, hereafter called a Corporate Audit Scheme, wherein multiple sites are assessed to be provided a single certificate, the following additional guidelines apply before a registrar can apply a Corporate certificate for QS-9000.

In order to adequately assess the quality system, it is necessary to visit every site but it is recognized that the number of man-days required to effectively assess each site may be less per site than the number given in the Appendix H chart.

The conditions required of the company for a Corporate certificate include:

a The quality system must be centrally structured and managed, and subjected to regular QS-9000 compliant internal audits at all sites.

b The quality system must comply with QS-9000/ISO 9001 or QS-9000/ISO 9002. If the system includes ISO 9001, all design activities must be evaluated.

c The balance of activities which <u>could</u> be centrally managed include:

1) contract review, where local acceptance of orders is permitted;
2) approval of suppliers;
3) evaluation of training needs (activity may have local aspects);
4) quality manual (Level 1 and Level 2) documentation and changes in same;
5) management review;
6) evaluation of corrective actions, but not necessarily implementation;
7) internal audit planning and evaluation of the result;
8) quality planning and continuous improvement activities (activity may have local aspects); and
9) design activities.

Note: Variations are acknowledged due to size and/or organizational structure.

The registrar must establish, during the quotation process, how the multi-site company falling under the Corporate scenario meets these requirements.

Man-Day Adjustment for Corporate Audit Scheme

<u>As a minimum, for a corporate certificate, the on-site audit man-days per site, are not expected to fall below 70% of the manday values per site shown in the Appendix H chart Survey Audit Days Table (as amended in 9503-R13). The same logic applies to the surveillance man-days in the Appendix H chart. Sites are defined as</u>

locations at which production processes occur; corporate schemes apply only to multiple site registrations. Remote locations, e.g., Engineering, Purchasing, must be audited as they support a site(s), but man-days to conduct these audits are included in a site audit as defined in the Appendix H Table.

Accreditation Body Notification

It is recognized that in Corporate multi-site audit approaches, the on-site audit man-days per site may justifiably be reduced to 70% of the levels shown in the Appendix H chart for on-site audit days and/or surveillances.

For any site approach used by a QS-9000 qualified registrar, if the registrar quotes man-days per site below the minimum levels shown in Appendix H, the registrar must notify its QS-9000 accreditation bodies of the quoted man-days via the QS-9000 Reporting Table. Also, he must provide the relevant supplier information, i.e., employees, number of sites, and product scope, in order to justify the quoting of fewer man-days than Appendix H minimums.

For any corporate approach used by a QS-9000 qualified registrar, if the registrar quotes man-days per site below 70% of the minimum levels per site shown in Appendix H, the registrar must submit/notify its QS-9000 accreditation bodies of the quoted man-days via the QS-9000 Reporting Table. Also he must provide the relevant supplier information, i.e., employees, number of sites, and product scope, in order to justify the quoting of fewer man-days than permitted.

These notifications must occur within five days of the quotation date to the client. The accreditation body is expected to review each of these inputs and take corrective and preventive action where appropriate.

QS-9000 audit proposals with suppliers involving violations of the current interpretation of Appendix H must be revised with those suppliers. This requirement for justification and notification of accreditation bodies applies to all registration audits occurring after 8/1/95.

Registrations issued prior to 8/1/95 must be brought into conformance with these interpretations of Appendix H over the next two surveillances.

Non-compliance places at risk the registrar, accreditation body and the resulting supplier QS-9000 certification.

Must a supplier certified to QS-9000:August 1994 be reassessed to the QS-9000:Feb 1995 version? How soon?

The February 1995 version of QS-9000 is now available from AIAG. QS-9000:1994 certified suppliers had until December 31, 1995 to be upgraded to the 1995 second edition of QS-9000. All QS-9000 certificates must include the date; i.e., QS-9000:1995. Registrars are encouraged to conduct any needed upgrade at the next 1995 certification surveillance. The upgrades are considered minor in magnitude.

If a customer tells a supplier they don't need to do a complete PPAP, what then is the requirement that must be certified to during an audit? -- the QS-9000? -- the PPAP? -- or the Customer?

The QS-9000 registration process certifies that the company is in compliance to the QS-9000 requirements, including PPAP. A customer can require only a Level 1 submission, but the PPAP file must be in place at the audit to determine compliance to PPAP and QS-9000 by the internal or second/third party auditors.

TRAINING (T)

Is the QSA (Quality System Assessment) Awareness Training Course sufficient to meet the training

requirements for plant personnel to become first party internal auditors, as required by element 4.17 in QS-9000?

No, the only sanctioned QSA Awareness Training Course is for general awareness; it alone does not fulfill element 4.17 requirements.

A) Is one of the QS-9000 qualified registrars, our competitor, recognized by IASG to provide training on QS-9000? B) What types of recognition do you give? C) Are there bodies recognized for qualifying QS-9000 registrar auditors (i.e., General Physics) and other bodies recognized to give general QS-9000 training (i.e., generic training for internal auditors, etc.) but not to qualify registrar auditors?

There is only one worldwide provider of QS-9000 registrar training recognized by the Big Three, General Physics Corporation (GPC). There are no registrars qualified to deliver registrar training.

The Big Three Task Force have selected one provider, BVQA, for delivery of the sanctioned QS-9000 supplier training for locations outside of North America. All providers of this training are subject to the restrictions of Appendix B and QS-9000 definition for consulting.

When and where are registrar auditor QS-9000 certification training courses being given in North America, Europe and elsewhere? I am a registrar in (Germany, Venezuela, Australia, Japan, United Kingdom) and need to equip my auditors.

Auditor QS-9000 certification training is available in many locations throughout the world. QS-9000 recognized registrars may schedule their auditors by calling AIAG at 810-358-3003 in the USA.

5 Conclusion

This chapter contains sufficient information about the world of QS-9000 for the reader to understand the 5 Ws and 1 H involved in planning the company's registration effort.

Chapter 3
Put It in Gear

The goals of any business are to provide customer satisfaction and to be profitable. It is important to recognize that to achieve the above goals, Man, Machine, Material, and Method (4 Ms) must operate in a proper Environment (E). This forms a working system of 4 Ms + 1 E.

In Figure 3.1 top management is represented by the guitarist. The quality system is depicted by the guitar consisting of the 4 Ms and E as denoted by the strings. The product is music.

The quality of music depends on the skill of management to guide and organize the activities of the system to achieve an effective combination of quality, cost, and delivery (QCD).

Management and the guitar are the inputs, music represents the output, customer reaction is the system feedback, and the net result is customer satisfaction and profitability.

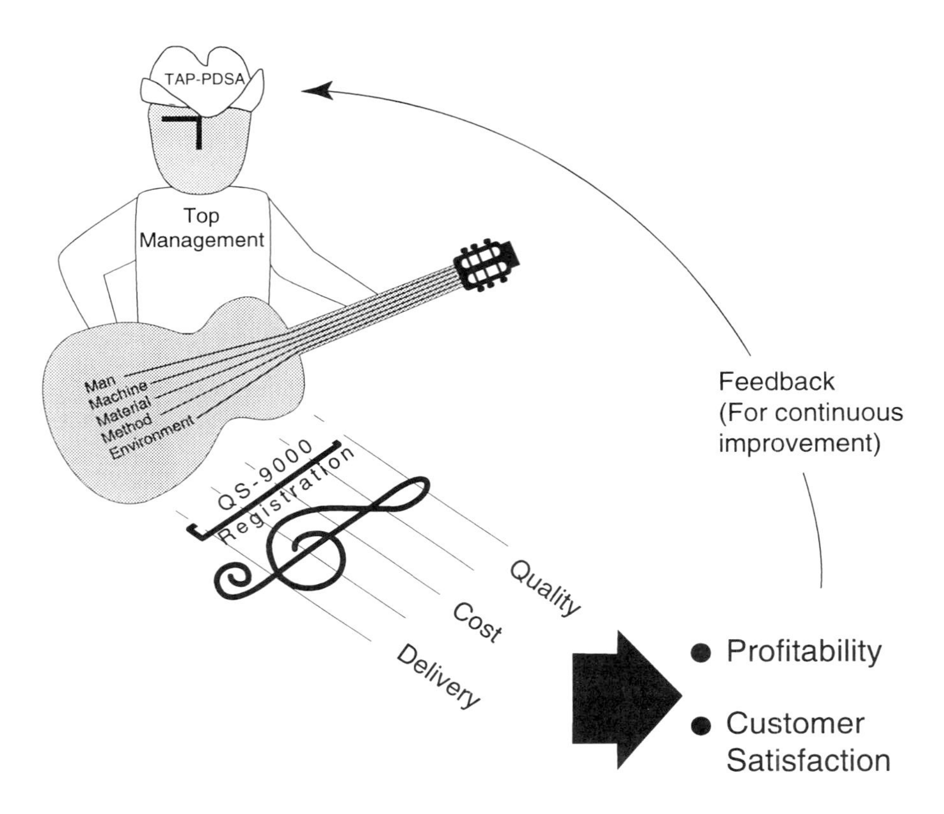

Figure 3.1 Guitarist Model of a Company

QS-9000 registration is an ideal way to orchestrate such a system into a quality-focused, customer-driven business able to compete in world-wide markets. However, to do so a proper framework must be established.

This chapter describes the framework required by addressing the following areas:

- Reasons for and Benefits of QS-9000 Registration
- Management Commitment
- Company-Wide Involvement
- Working Environment

1 Reasons for and Benefits of QS-9000 Registration

1.1 Reasons for QS-9000 Registration

Every company has its own reason for pursuing registration. The most common reasons are:

- Customer Requirement
- Marketing Strategy

1.1.1 Customer Requirement

Many automotive companies require QS-9000 registration of their supplier's quality system as a condition for conducting business. In such a scenario, QS-9000 registration becomes a requirement and the supplier is left with no other option but to pursue registration. Table 3.1 lists the supplier registration deadlines for Chrysler, Ford, and GM.

Table 3.1: QS-9000 Third Party Registration Deadlines

Company	Registration Deadlines
Chrysler	July 31, 1997
Ford	Not required presently
GM	December 31, 1997

1.1.2 Marketing Strategy

Obtaining registration can be a part of a company's marketing strategy and provides an excellent sales pitch. The QS-9000 requirement is recognized almost globally by the automotive industry, and thus registration provides international acceptance. The Big Three and other global markets open and become more accessible, providing business growth opportunities. Registration also provides an edge over the competition if they are not registered and a leveling of the field if they are.

1.2 Benefits of Registration

Besides fulfilling the specific reasons, there are other benefits acquired as a result of registration:

- Continuous Improvement
- Improved System Documentation
- Improved Customer Relationships
- Improved Quality Awareness

Note: It is important to remember that all benefits of registration will not be obtained overnight. Some of them will be realized immediately and others will accrue slowly and become fully evident only after a period of time.

1.2.1 Continuous Improvement

Registration requires the development of a quality system which becomes an excellent starting point for continuous improvement. Continuous improvement involves company-wide activities which aim at improving customer satisfaction.

1.2.2 Improved System Documentation

The QS-9000 registration effort provides an excellent opportunity to eliminate excess documentation. This effort results in a lean and proactive quality system that contains better documents and is usually more efficient and cost effective.

1.2.3 Improved Customer Relationships

Another registration benefit is usually "an improved customer–supplier relationship." The customer has confidence that the supplier is quality conscious and interested in supporting the customer through continuous improvement. This, in turn, may result in fewer customer requests (i.e., QC data, tests, etc.) and audits.

1.2.4 Improved Quality Awareness

The company that practices quality soon becomes aware of the fact that good quality pays a variety of dividends. For example, marketing appreciates that manufacturing can meet its commitment to quality, price, and delivery and seeks opportunities to utilize these newfound strengths. Customers return to suppliers of products that they are satisfied with. Engineering increases reliability of design, knowing that operations will consistently execute these higher requirements. Top quality companies attract the best people available. Success builds on success!

2 Management Commitment

Most projects that fail do so due to lack of management commitment and direction. This is also true for the QS-9000 registration project.

The importance of management guidance and participation is an integral part of the quality system requirements (4.1 Management Responsibility)

ISO 9004-1 states: "The responsibility for, and the commitment to a quality policy belongs to the highest level of management. Quality management encompasses all activities of the overall management function that determine the quality policy, objectives and responsibilities, and implement them by means such as quality planning, quality control, quality assurance, and quality improvement within the quality system."

Management must understand the scope of the registration effort and that it requires a fair amount of financial and human resources.

Top management must evaluate the above and then make the decision to proceed with QS-9000 registration.

3 Company-Wide Involvement

Every employee in any organization is involved in some way or other in the registration effort. The common goal is to obtain QS-9000 registration. Only visible management support and guidance can make this happen.

3.1 Top Management

Top management defines its policy for quality, objectives for quality, and its commitment to quality (Section 4.2, Quality System).

3.2 Middle Management

Middle management ensures that the policies and objectives of upper management are communicated and implemented down into the organization through system procedures.

3.3 Technical Staff

The technical staff develops and documents product specifications and their manufacturing processes.

3.4 Supervisors

Supervisors develop area-specific instructions and ensure that these instructions are understood and implemented.

3.5 Front-Line Employees

Front-line workers make up the majority of the workforce, and it is they who manage and operate the processes by following quality documentation and work instructions.

4 Working Environment

Imagine people walking around with gloomy faces, unhappy with their work, blaming each other for mistakes, fighting problems day in and day out, and focusing on quantity not quality. Productivity will slip considerably and morale is bound to decline. This should not occur. It is important that management remain diligent about establishing and maintaining a good working environment. People should have pride in their work and feel that they are contributing to the "wellness" of the company.

It is obvious that a proper environment is crucial to the success of a project. QS-9000 registration is no exception. For the registration project to be successful in the long term, it will require that all personnel work together in harmony to achieve the common goal of registration. The creation of the environment is best explained by Dr. Deming in his 14 points of management.

Deming states: "The fourteen points all have one aim, to make it possible for people to work with joy."

The points consist of simple statements requiring deeper understanding. Their implementation requires a cultural change of focus from short-term profitability to long-term "constancy of purpose for improvement of product and service" in most companies.

Dr. Deming explains each point in his own words (Courtesy Wootten Productions), and Pulitzer Prize-winning cartoonist Pat Oliphant illustrates the points through his sketches.

4.1 Deming's 14 Points: A Philosophy of Life

Point #1: Create constancy of purpose for the improvement of product and service.

Pat Oliphant Illustration (Permission: Wootten Productions)

Figure 3.2 Constancy of Purpose

Point #2: Learn the new philosophy. Teach it to employees, to customers, to suppliers. Put it into practice, in other words – the new philosophy – which is one of cooperation, win-win, everybody wins.

Pat Oliphant Illustration (Permission: Wootten Productions)

Figure 3.3 Everybody Wins

Point #3: Cease dependence on mass inspection. Much better to improve the process in the first place so that we don't produce so many defective items – or none at all.

Pat Oliphant Illustration (Permission: Wootten Productions)

Figure 3.4 Design Quality In

Point #4: End the practice of awarding business on the basis of price tag alone. Instead, minimize total cost in the long run. That means one has to predict the cost of use on any product or service.

Pat Oliphant Illustration (Permission: Wootten Productions)

Figure 3.5 Don't Buy on Price Tag Alone

Point #5: Improve constantly every process for planning, production, service, whatever the activity is.

Pat Oliphant Illustration (Permission: Wootten Productions)

Figure 3.6 Continuous Improvement

Point #6: Institute training for skills. People learn in different ways, and training must take account of those differences.

Pat Oliphant Illustration (Permission: Wootten Productions)

Figure 3.7 Training for Skills

Point #7: Adopt and institute principles for the management of people. I'm referring to the management of people for recognition of different abilities, capabilities, aspirations.

Pat Oliphant Illustration (Permission: Wootten Productions)

Figure 3.8 Institute Leadership

Point #8: Drive out fear, build trust. It's purely a matter of management.

Pat Oliphant Illustration (Permission: Wootten Productions)

Figure 3.9 Drive Out Fear

Point #9: Break down barriers between staff areas. In other words, build a system. Build a system within your organization for win-win, where everybody wins. This means cooperation. It means abolishment of competition.

Pat Oliphant Illustration (Permission: Wootten Productions)

Figure 3.10 Break Down Barriers

Point #10: Eliminate slogans, exhortations, targets asking for zero defects, new levels of productivity. Nonsense! If you don't need a method, why weren't you doing it last year? Only one possible answer: You were goofing off.

Pat Oliphant Illustration (Permission: Wootten Productions)

Figure 3.11 Eliminate Slogans

Point #11: Eliminate numerical goals, numerical quotas for anybody. A numerical goal or quota accomplishes nothing.

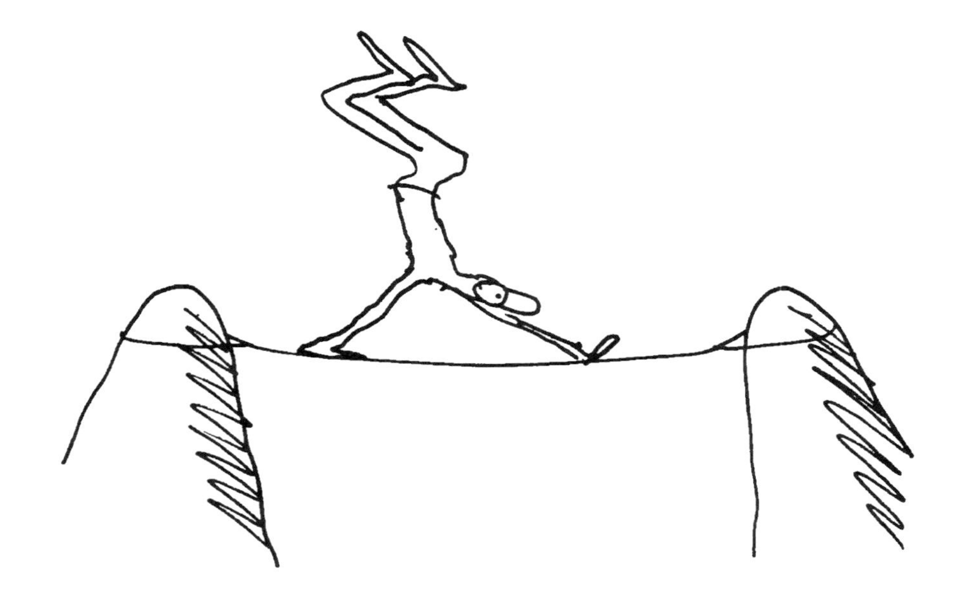

Pat Oliphant Illustration (Permission: Wootten Productions)

Figure 3.12 Method

Point #12: Remove barriers that rob people of joy in their work. This will mean abolishing the annual rating or merit system which ranks people, creates competition and conflict.

Pat Oliphant Illustration (Permission: Wootten Productions)

Figure 3.13 Joy In Work

Point #13: Institute a vigorous program of education and self-improvement.

Pat Oliphant Illustration (Permission: Wootten Productions)

Figure 3.14 Continuing Education

Point #14: Accomplish the transformation; that is, continue to study the new philosophy. Develop a critical mass in your organization that will bring about the transformation.

Pat Oliphant Illustration (Permission: Wootten Productions)

Figure 3.15 Accomplish the Transformation

5 Conclusion

This chapter represents sound business thinking! It explains that three elements – a reason for QS-9000 registration, top management commitment and company-wide involvement, and a conducive working environment – are essential for a successful long-term registration.

These elements represent the "Get-Ready" stage for the registration effort, and the next chapter represents the "Go" stage!

Chapter 4
Step on the Gas

This chapter provides the reader with the necessary first steps to get a QS-9000 registration effort "jump started."

Once top management has given a go-ahead for the registration effort, it is time to proceed. It is important that the project be properly launched as organizations waste a significant amount of time and effort trying to find the right direction. "Jump starting" prevents this situation from happening and results in a shorter overall completion time. "Jump starting" consists of the following important activities:

- Selection of a Management Representative
- Formation of a Project Team
- Needs Assessment Audit (NAA)

- Preliminary QS-9000 Registration Plan
- Registrar Selection
- Measurements System
- Quality System Documentation (QSD)

The TAP approach is demonstrated here once again. Selection of the management representative, formation of the project team, and their training represents the "T," the needs assessment audit represents the "A," and preparation of the preliminary registration plan represents the "P."

It is important to point out that this chapter provides only the basic information regarding jump-start activities. Detail is provided in later chapters.

Chapter 5, *The Drive (TAP)*, covers QS-9000 training requirements, various audits, and registration plan.

Chapter 6, *Driving Between the Lines,* covers measurements and costs.

Chapter 7, *Registrar Selection*, presents the registrar selection process.

Chapter 8, *The Owner's Manual*, explains the design, documentation, and implementation of the quality system.

1 Selection of a Management Representative

After the decision to seek registration is made, the first important activity is the selection of a management representative by executive management.

1.1 What Is a Management Representative?

Using the words of the standard, a management representative is a member of management who is appointed by the supplier's executive management to:

- Assure that the quality system requirements are established, implemented, and maintained in accordance with the International Standard.
- Report on the performance of the quality system to the supplier's management for review and as a basis for improvement of the quality system.

Note: The responsibility of a management representative may also include liaison with external bodies on matters relating to the supplier's quality system.

The management representative is usually the individual who leads the registration effort and heads the project team.

It should be recognized that the successful management representative takes on an assignment that will probably become a significant, if not total, allocation of time. If the commitment is not recognized, the registration effort will be seriously impaired.

1.2 What Are the Duties of a Management Representative (MR)?

It is recommended that a position description be prepared for the management representative. The responsibilities of the MR should be clearly defined and documented to enable the MR to perform his assignment effectively. A model position description (duties, skills, knowledge, and training) for the management representative/project leader is provided in Table 4.1.

Table 4.1: Management Representative Matrix

Duties
• Ensure that quality system requirements are established, implemented, and maintained in accordance with the QS-9000 requirement. • Report on the performance of the quality system to the supplier's management for review and as a basis for improvement of the quality system. • Lead the registration effort. Select and lead the project team. • Be responsible for the internal audit program. • Develop quality system documentation, implementation approach, and strategy. • Determine and provide for company-wide training. • Manage registrar selection and interface process. • Develop project cost and measurement criteria. • Maintain the registration as needed throughout the life of the registration.
Skills and Knowledge
• Project management and leadership skills. • Internal auditing skills. • Documentation skills (in-depth understanding of the family of QS-9000 Standards). • Training skills. • Registrar knowledge. • Accounting skills.
Training
• QS-9000/QSA • PPAP & FMEA • APQP & Control Plan • Lead assessor training

1.3 Who Should Be the Management Representative?

It is ideal for the QS-9000 project manager to be the management representative. In most companies the quality assurance manager usually assumes the responsibility of a management representative.

2 Formation of a Project Team

Having selected the management representative, the next task is to assemble a project team for the purpose of obtaining registration.

2.1 What Is a QS-9000 Project Team?

A QS-9000 project team is a group of professionals representing all of the functional departments (engineering, quality, documentation, manufacturing, marketing, materials, information systems) that are within the scope of the QS-9000 registration and are charged with developing an QS-9000 compliant quality system.

Note: It is advisable to select people who want to be on the team.

2.2 What Are the Duties of the Project Team?

The duties of the project team are quite similar to the duties of the project leader. Each project member has some area of specialization required for the QS-9000 project. Table 4.2 provides a matrix of duties, skills, and training requirements for the project team.

Note: Most companies jump straight to the Needs Assessment Audit and pay little attention to the training needs of management representative/project leader and the project team. Experience has shown that an initial investment in training the project team will help speed up the registration process significantly.

Table 4.2: Project Team Matrix

Duties
• Work together with project leader to achieve registration. • Perform internal audits. • Develop quality system documentation and its implementation. • Determine and provide for company wide QS-9000 training. • Assist in registrar selection. • Develop and perform project measurements and costing. • Help maintain the registration throughout the life of the registration.
Skills and Knowledge
• Project Management/Team Management Skills. • Internal Auditing Skills. • Documentation skills (In-depth understanding of the QS-9000 quality system requirements). • Training Skills. • Accounting Skills.
Training
QS-9000 Implementation Training includes: • QS-9000/QSA • PPAP & FMEA • APQP & Control Plan • QS-9000 Internal Auditor Training • QS-9000 Documentation Training

2.3 Why Do We Need a Project Team?

QS-9000 implementation is not a one-man show. QS-9000 is a multifaceted, company-wide project involving all departments. To efficiently achieve registration, representatives from each department must work together as a team towards the common goal of registration.

3 Needs Assessment Audit (NAA)

When the management representative and the project team (trained auditors) have been trained, it is time to perform the Analysis step (the A in TAP), or the needs assessment audit.

The NAA results are a major decision point for management. At this time, management receives another opportunity to reconfirm its commitment to the registration effort.

3.1 What Is a Needs Assessment Audit?

The first comprehensive internal audit per the QS-9000 requirement will be a needs assessment audit. The NAA examines existing documentation and its implementation.

This audit is primarily performed to provide a clear, in-depth picture of the deficiencies within the existent quality system so that proper planning (P in TAP) can be done to bring the quality system into compliance.

In addition, this information is also extremely vital in determining the resources required to achieve registration (i.e., cost/budget and man-hours).

The results of the NAA must be properly communicated to all affected. This ensures that everyone is focused in the same direction. Often a formal meeting, conducted by senior management, is an effective vehicle to communicate these concerns.

The NAA itself can be used not only to audit, but also to train at the same time. This approach is very useful as it reduces the training time and standardizes the QS-9000 interpretation.

3.2 What Should You Audit?

Audit the complete quality system activities that are within the scope of the QS-9000 quality system requirements.

Note: It should be noted that almost all departments will be affected by the scope of registration and will need to be audited.

3.3 Who Performs the Needs Assessment Audit?

The NAA is an assignment for the project/internal audit team.

It is acceptable to have the needs assessment audit performed by external auditors (i.e., consultants or contract personnel), however, auditing one's own system gives a better understanding of how things really stand!

A conflict of interest would prohibit the registrar candidates from participating in this activity.

3.4 How Is a Needs Assessment Audit Performed?

The trained auditor uses a systematic approach as shown in Figure 4.1.

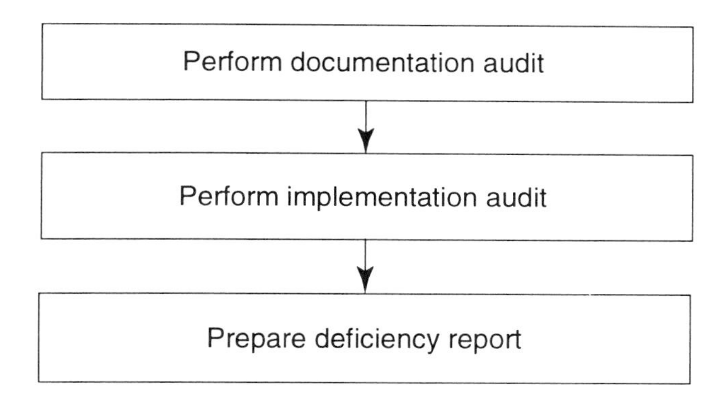

Figure 4.1 Needs Assessment Audit Flowchart

Perform Documentation Audit

Perform an audit of the documentation utilizing the checklist provided in the Quality System Assessment Manual detailed in question 1 of Element 4.2. In the audit, follow the hierarchy of documentation. (For example, a work instruction must implement a procedure. The procedure in turn must implement a quality policy.) The key questions are: "Have you documented what you do?" and "How does it comply to the standard?" After auditing all requirements, determine the documentation compliance level.

Perform Implementation Audit

The auditor then observes the element being performed utilizing the Quality System Assessment Manual. The key question is: "Have you implemented what has been documented?" After auditing all the requirements for implementation, determine the implementation compliance level.

Prepare Deficiency Report

Prepare a detailed deficiency report containing the overall compliance level. This report serves as an excellent baseline for measuring project progress.

4 Preliminary QS-9000 Project Plan

After the completion of the NAA (A in TAP), the preliminary project plan (P in TAP) is developed.

4.1 What Is a Preliminary QS-9000 Registration Plan?

A preliminary QS-9000 project plan identifies major milestones and their expected completion dates as shown in Figure 4.2.

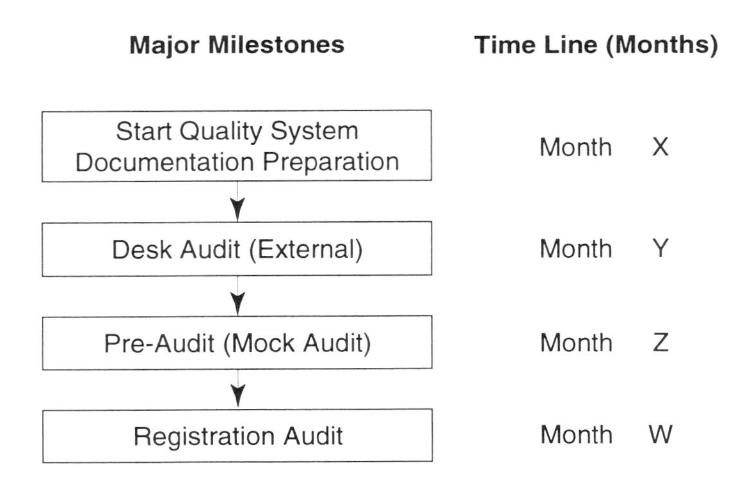

Figure 4.2 Preliminary QS-9000 Project Plan

Start Quality System Documentation Preparation

This step involves the development and documentation of a QS-9000 system.

Desk Audit (External)

At this point, the registrar examines the documentation against the QS-9000 requirements.

Pre-Audit (Mock Audit)

This audit is usually a preparatory audit used to determine "system readiness" and correct any last-minute deficiencies. It is usually an optional activity.

Registration Audit

The registration audit is the moment of truth when the registrar performs a formal compliance audit.

Note: When developing the time line for the above flowchart, the following key point must be considered:

- It is important to block out dates with the registrar. Many companies have had to delay registration as registrars were not available to perform audits when desired.

5 Registrar Selection

The selection of a registrar should be an early priority for the management representative and the project team.

It should be noted that a long-term relationship between the registrar and the company should be developed since this relationship will exist as long as the company wishes to be registered.

5.1 Why Is Early Registrar Selection Important to the Registration Process?

- The registrar selection process becomes a learning experience since it involves dealing directly with the registrar who is thoroughly familiar with the registration process.
- Registrars are not permitted to consult, but they can often provide guidance simply by telling about their procedures and how they operate. (One of the skills that team members who work on the registrar selection process develop is how to phrase a question for information that the registrar's representative can respond to without tripping over the consulting issue.)
- The project team starts to learn about "delivery" and "costs" when the registrar discusses response time to audit requests and the costs related to these activities.

5.2 Who Will Select the Third Party Registrar?

In most companies, the registrar is selected by the management representative and project team. Consensus is strongly recommended.

5.3 What Are the Steps in Selecting a Registrar?

Figure 4.3 outlines the major steps for selecting a registrar.

Determine What Registrars Are Available

To locate third party QS-9000 approved registrars in the U.S. contact:

Registrar Accreditation Board
c/o American Society for Quality Control
611 East Wisconsin Avenue
Milwaukee, WI 53201

Phone: 800-248-1946
Fax: 414-272-1734

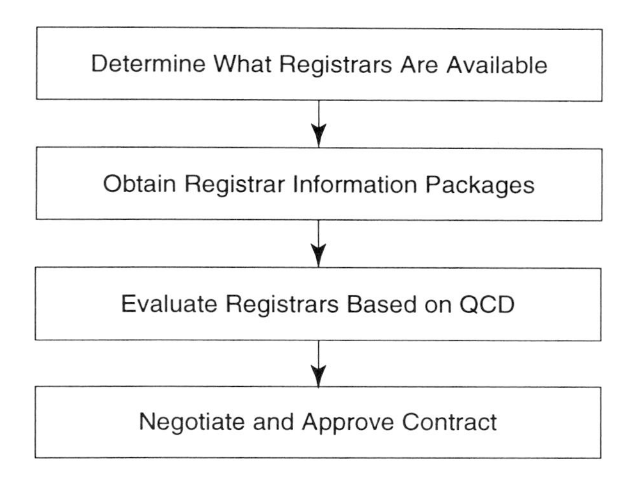

Figure 4.3 Registrar Selection

Obtain Registrar Information

Contact each registrar and request an information package, listing companies that the registrar has registered. These packages should be reviewed and evaluated carefully by the project team. They are the basis for the first screening out of unsuitable candidates.

Evaluate Registrars

It is strongly recommended that the sub-committee for registrar selection prepare a comprehensive checklist or evaluation guide based upon QCD principles. The detailed evaluation procedure is covered thoroughly in Chapter 6. The project team needs to winnow the list of third party registrar candidates down to approximately three finalists from whom the project team will select the registrar to work with.

Negotiate and Approve Contract

The project team approves the registrar candidate and initiates the negotiation process. The management representative is usually assigned this responsibility. The registrar is asked to submit a contract for approval. Upon receipt the project team evaluates the proposed contract to ensure that the all requirements have been satisfied. If the contract is satisfactory it is approved.

6 Measurement System

All project plans should have a measurement system built into them. In the words of Peter Drucker, "If you cannot measure it, then you cannot manage it."

6.1 What Are Measurements?

Measurements are **analyses** in TAP resulting in an improved plan and **study** in PDSA resulting in appropriate corrective action.

6.2 QS-9000 Project Measurements

Project measurements are measurements that are used to track the QS-9000 project and are based on the NAA results. NAA results provide a clear picture of the system's compliance level at the start of the project against which the planned progress can be measured and tracked.

Typical QS-9000 project measurement elements are:

- Activities against Time, as shown in the Gantt Chart (plan section in Chapter 5) (documentation compliance levels, audits, observations, and respective corrective actions, etc.)
- Cost (money spent against budget)

6.3 Quality System Performance Measurements

Performance measurements are measurements that determine the improvements in quality system performance that resulted due to the registration.

This type of measurement consists of the "key" system quality, cost, and delivery (QCD) elements indicating the health of the system. Identify these elements at the start of the project so that improvements in the quality system that resulted due to the successful registration effort can be measured.

Typical quality system performance measurement are:

- Customer returns
- Reject ratios
- Process capability indices

The project and system measurements are best determined using a matrix of 4 M's and QCD as explained in Chapter 7.

7 Quality System Documentation

The preparation of quality system documentation (QSD) that is to be in compliance with the QS-9000 requirements and its effective implementation is the heart of the registration effort. QS-9000 states the following:

"The Supplier shall prepare documented procedures consistent with the requirements of this international standard and effectively implement the quality system and its documented procedures."

7.1 What Is Quality System Documentation?

It is documentation that describes an organization's quality system.

ISO 8402 defines "**quality system**" as "The organizational structure, responsibilities, procedures, processes, and resources for implementing quality management." A **document** is anything that is written.

The QS-9000 standards state the following:

"The supplier shall establish, document, and maintain a quality system as a means of ensuring that product conforms to specified requirements. The supplier shall prepare a quality manual covering the requirements of the International Standard."

7.2 How to Prepare Quality System Documentation

Quality system documentation is developed in two steps.

- Designing the quality system
- Documenting the quality system

Designing the Quality System

Quality system design involves mapping or flowcharting all quality system activities from customer input, product design, through development, production, installation, and servicing.

Documenting the Quality System

Quality system documentation involves documenting the above. ISO guidance standard 10013, Guideline for Developing Quality Manuals, provides excellent direction on documentation organization and preparation.

Note: The QSD process is explained in detail in Chapter 8.

8 Conclusion

This chapter identifies and explains the first essential steps towards an efficient registration effort using the TAP approach. The information provided is drawn from real experiences and should help launch the registration effort in the right direction. Further detail on these topics is provided in the next four chapters.

Chapter 5
The Drive (Training, Audits, Planning)

This chapter presents the training, auditing, and planning (TAP) requirements in detail for the QS-9000 registration effort. The **T**raining element details the training requirements for all levels of workforce, the **A**nalysis element explains the various audits required during the registration process, and the **P**lan explains the route to registration through the development of a TAP-PDSA roadmap.

1 Training

Training should be the first step in any registration effort and represents both training, which can be implemented immediately, and education, which provides the intellectual insight to understand and appreciate the entire quality management system, now and in the future.

In the TAP approach, training is multifaceted and covers the entire workforce from executive management to the front-line workers.

The approach described is based on the idea that a successful long-term QS-9000 registration begins and ends with training. Since registration is a continuous process, training must also be a continuous process. However, the training needs must be planned and defined well in advance.

1.1 QS-9000 Project Training Matrix

The QS-9000 project training matrix (Table 5.1) lists the typical training requirements for an organization to achieve QS-9000 registration. The familiar 5 W approach is used.

Who Who defines the individuals and groups who are to receive the training.

What What describes the course work or topic.

Why Why states the purpose and anticipated results of training.

Where Where addresses the source of training (university/ consultant/ books, etc.).

When When indicates the timing when training should be conducted.

The matrix is generic. It is recommended that the organization seeking registration compile a similar training matrix to meet their specific requirements. The cost estimates must also be made for budgeting purposes.

1.2 Executive Management Training

1.2.1 Who? Top Management Policy Makers

Executive management training is for top management policy makers such as CEO, COO, vice presidents, and business unit managers.

Table 5.1: QS-9000 Project Training Matrix

Who	What	Why	Where	When	Cost
Executive Management	QS-9000 Executive Overview	Provides executive awareness, commitment and participation	External	First step	$500
Management Representative (MR)	Lead Auditor Training	Provides a comprehensive understanding of QS-9000 registration program	External	After management commits to the project	$1500
Project Team (incl. MR)	QS-9000/QSA APQP and CP PPAP and FMEA QS-9000 Documentation	Provides "know-how" to implement registration project	External	After formation of project team	$400 $200 $400 $450
Internal Audit Team	QS-9000 Internal Auditor Training	Provides skills required for internal auditing	External	Prior to needs assessment audit	$450
Managers, Supervisors, Engineers	QS-9000 Overview	Provides direction for project implementation	Internal	Before documenting quality system	Internal Costs
Front-Line Workers	Quality Policy, Job Skills, and Audit Survival Training	Implement system	Internal	Ongoing	Internal Costs

1.2.2 What? QS-9000 Executive Overview

The following are typical topics to be covered in a "QS-9000 executive overview" class.

- Comprehensive overview of the background and development of the QS-9000 quality system requirement
- Introduction to the QS-9000 quality system requirement
- The need and importance of executive involvement in an effective Quality Management System
- Business reasons/strategy for seeking QS-9000 registration
- Registration economics
- Insight into the registration process
- QS-9000 as a tool for continuous improvement

1.2.3 Why? Provides Executive Awareness

Executive management training introduces top management to the advantages, costs, and resources involved with the registration effort. This information is necessary for management before it commits to the registration effort. It also represents the "voice from the outside" which is important to strengthen this commitment.

Top management provides the leadership for any registration effort. These managers need to appreciate the impact of a QS-9000 registration effort on the quality strategy and the corporation's business plan.

1.2.4 Where? Externally

Top management training can come from many sources such as educational institutes, customers, and seminars.

1.2.5 When? First Step

Top management education must start well in advance of the QS-9000 registration effort.

1.2.6 Cost? $500

A one-day seminar for senior management presented by a "name" provider will cost about $500 per person.

1.3 Management Representative Training

1.3.1 Who? Management Representative

It is recommended that the management representative (MR) take an approved QS-9000 Lead Auditor course. The MR should also consider taking the courses listed for the project team.

1.3.2 What? Lead Auditor Training

This comprehensive, Lead Auditor Training, 36-hour course is designed to provide a thorough understanding and knowledge of auditing in accordance with the ISO 10000 series of standards. It focuses on intensive case history studies. The material covers additional topics, such as the registrar's auditing process, documentation structures, and systems. This training involves working in teams and provides a chance to conduct a simulated third party audit. Trainees are also required to pass a final examination on the 5th day for full qualification.

Table 5.2 shows a typical approved Lead Auditor course outline.

Table 5.2: Lead Auditor Course Outline

Day	Key Session Topics
Monday	• Introduction to the assessment process • Planning & preparation • Scopes of assessment • Case study: Scopes • Managing the assessment
Tuesday	• Individual tutorials • Document review • QS-9000 interpretation • Case study: Review of documentation against QS-9000 • Recording the document review • Exercise: NCN writing
Wednesday	• Individual tutorials • Case study: Prepare assessment plan • Approaches to conducting an assessment • Assessment process • Case study: Prepare for and hold an opening meeting • Exercise: Prepare an Aide Memoire • Gathering information
Thursday	• Conducting the audit • Audit reporting (closing meeting) • Team presentations of case-study findings • QS-9000–a global perspective • Auditor Certification process in the U.S.
Friday	• Review of exam • Examination (2 hours)

Source: LRQA, Hoboken, NJ

1.3.3 Why? Provides Comprehensive Understanding of QS-9000 Registration Program

In order to have a successful registration effort, those selected for this task need to be the most highly and comprehensively trained in the whole body of QS-9000 and ISO 10000 quality standards, their purpose, and application. They must be the "voice of authority." The purpose (why?) of this training has the strongest specifics.

In short, this course provides the following information, which is critical for the registration effort:

- In-depth understanding of the family of QS-9000 quality systems requirements standards.
- Understanding of the mechanics of the registration process.
- Selection requirements for those registrars which are most beneficial to the individual company business plan.
- Credentials to deal effectively with the registrar, senior management, and customers.
- Ability to manage and conduct effective internal audits of the quality system components to the requirements of QS-9000 requirement.

1.3.4 Where? Externally

The lead assessor training must be undertaken externally. These courses are generally offered by approved providers at locations globally. Since the number of courses offered continues to grow, it is usually possible to locate a time and place convenient to the individuals to be trained. Since the training format involves working in teams, it is often beneficial to undertake this training as a group when more than one person is selected to receive lead assessor training.

1.3.5 When? After Management Commits to the Registration Project

Lead assessor training should be undertaken as soon the decision to seek a QS-9000 registration is made.

1.3.6 Cost? $1500

Costs vary from $1000 to $1500 per person (USD). For companies with a limited budget the two-day internal auditor course should be adequate.

1.4 Project Team Training

1.4.1 Who? Project Teams

The project team, consisting of representatives of each department who will develop and implement the registration plan, will need to be trained.

1.4.2 What? QS-9000 Project Training

- QS-9000/QSA Training
- APQP/CP Training
- PPAP and FMEA Training
- QS-9000 Documentation Training

These in-depth courses provide a thorough understanding of QS-9000 requirements, the related reference manuals, and their documentation requirements.

Tables 5.3 through Table 5.7 show typical outlines for these courses.

Table 5.3: QS-9000/QSA Course Outline (2 days)

Key Session Topics
• Goal and purpose of QS-9000
• The approach to using QS-9000 and who it applies to
• The 23 QS-9000 elements (automotive interpretations)
• Contents of the customer-specific sections
• The quality system development/implementation
• Flow of QS-9000 process (first, second, and third party assessment processes)
• Code of practice of quality system registrars
• Supplier responsibility regarding third party registration

Source: AIAG, Detroit, MI

Table 5.4: APQP/CP Course Outline (1 day)

Key Session Topics
• Overview of tools, procedures, and reporting requirements of APQP/CP
• Explanations of activities involved in APQP
• Explanation of commonized and customer-specific requirements
• Example of Control Plan

Source: AIAG, Detroit, MI

Table 5.5: PPAP Course Outline (1/2 day)

Key Session Topics
• Overview of PPAP (scope, definition, and purpose)
• When submission is required
• Specific requirements for part approval
• Records and sample retention
• Part submission status
• Example of part submission warrants and associated forms

Source: AIAG, Detroit, MI

Table 5.6: FMEA Course Outline (1/2 day)

Key Session Topics
• Overview of FMEA (scope, definition, and purpose)
• Using FMEA in design and manufacturing processes
• Review of reporting requirements and associated forms
• Evaluation criteria
• Development of recommended actions and follow-up
• Examples of design and process FMEAs

Source: AIAG, Detroit, MI

Table 5.7: QS-9000 Documentation Typical Course Outline

Key Session Topics
• IS0 9000 and QS-9000 quality system requirements
• Registration process and QS-9000 Code of Practice
• Documentation requirements (purpose/structure/content)
• Guidelines for writing effective documentation
• Implementing and maintaining effective document control
• QS-9000 element-by-element explanation
• Quality system assessment approach

Source: SRI, Wexford, PA

1.4.3 Why? Provides QS-9000 Project Know-How

These courses provide information on how to implement QS-9000 key quality system components in an organization for effective registration.

- **QS-9000/QSA**

This class provides employees information regarding all the elements of QS-9000 and the various assessment processes.

- **APQP/CP**

This class provides employees information regarding product quality planning processes.

- **PPAP**

This class provides employees information required for product submission and approval processes.

- **FMEA**

This class provides employees of supplier companies information required for developing FMEAs.

- **QS-9000 Document Control Training**

Courses of this type provide a clear understanding of the QS-9000 documentation requirements and implementation strategies.

1.4.4 Where? Externally

It is recommended that the initial training be provided by outside providers unless there is a strong internal training group available.

1.4.5 When? After Formation of the Project Team

One of the key activities in implementing the registration process is selection of the project team members. Each member selected should have the individual's training needs carefully assessed by the management representative or the project lead. Often training can be initiated as team members are selected. All members should be trained before the team gets down to serious work. Under normal circumstances, training should be completed for this group within two months of the decision to proceed with the QS-9000 registration effort.

1.4.6 Cost? $99 to $500

The QS-9000/QSA training costs $400, APQP/CP costs about $200, PPAP and FMEA classes run up to $200 each, while the document control class costs approximately $450.

1.5 Internal Audit Team Training

1.5.1 Who? Internal Audit Team

These auditors are responsible for managing and performing internal and subcontractor audits for the organization.

The number of internal auditors required depends on the size of the company. However, it is recommended that a company should have a minimum of two auditors.

1.5.2 What? QS-9000 Internal Auditor Training

This two day comprehensive internal auditor class meets the training requirements for the internal auditor. It includes audit supervision and management, detailed audit workshops, and audit simulations. Table 5.8 shows a typical internal auditor course outline.

Table 5.8: Internal Auditor Training Course Outline (2 day)

Key Session Topics
• Coordinate an effective Quality System audit against QS-9000. – Defining audit scope and determining activities/areas to be observed – Establishing expectation, criteria and supporting checklists – Developing detailed audit plans and selecting audit team – Utilizing balanced audit methods and projecting a professional image – Applying effective auditor techniques such as observation, probing, confrontation, judgement – Conducting productive pre-audit and post-audit conferences – Reporting audit results: guidelines for report structure and content – Recognizing non-conformity and issuing corrective action notification

Source: SRI, Wexford, PA

1.5.3 Why? Provides Internal Auditing Skills

This course trains personnel to become auditors with the necessary skills to evaluate quality system compliance to the requirements of QS-9000 as required by Element 4.17.

Auditing is one of the most important quality system measurement methods. The results of the first internal audit, better known as the needs assessment audit (NAA), provide the system compliance level and determine the scope of the registration project. A poorly conducted needs assessment can send the registration effort in a completely wrong direction. Trained internal auditors will prevent such a tragedy.

Secondly, the initial purpose of the internal auditors is to bring the quality system into QS-9000 compliance through a series of internal audits and their corrective actions.

The internal audits are the most important short- and long-term continuous improvement tools.

1.5.4 Where? Externally

Initially internal auditors are trained using external training. Later on, consider an internal auditor training program. Try having at least one auditor per department. This eliminates the need for full-time auditors (also auditors cannot audit their own areas). It then becomes possible to rotate auditors. This approach provides a fresh look each time; designated auditors tend to get a bad reputation.

1.5.5 When? Prior to the NAA

This audit should be conducted before the needs assessment audit.

1.5.6 Cost? $450

A two-day internal audit seminar presented by a "name" provider will run about $450 per person.

1.6 Middle Management

1.6.1 Who? Managers, Supervisors, and Engineers

These employees are responsible for managing the QS-9000 project implementation in their respective functional groups.

1.6.2 What? QS-9000 Overview

Overview training is similar to an executive overview class, with more emphasis on area-specific documentation requirements and audit preparation and less focus on the business issues.

Shown below is a typical course outline for an engineering group.

- Introduction to QS-9000 series of quality system standards (purpose, benefits, etc.)
- The registration process (desk audit, registration audit, etc.)
- Understanding QS-9000 requirements in detail

1.6.3 Why? Provide Direction for Documentation

These professionals are the core group and quality system success largely depends on their involvement. They control the process and will be responsible for documentation preparation and implementation. Most of the compliance problems will be dropped in their laps.

1.6.4 Where? Internally

Most training for this group of contributors is performed internally because of the close relationship to the actual activities that occur.

1.6.5 When?

Complete this training before the start of the quality system documentation process.

1.6.6 How? In Groups

Train specific groups together (i.e., train the engineering group together on requirements of QS-9000 that are applicable to them).

Use the "everyone receives training concept." It is important that project team members start to impart their knowledge to the functional groups as soon as possible. This process enables the quality culture to flow down and creates an environment of common cause. A shift in attitudes, which is literally a new paradigm, is important. A QS-9000 registration effort is often a difficult path to accomplishment, and a sense of togetherness is vital towards smoothing out the bumps in the road.

1.6.7 Cost? Internal Cost

This cost is calculated per the internal training program.

1.7 Front-Line Worker Training

1.7.1 Who? Front-Line Workers

Front-line workers are machine operators, assemblers, technicians, inspectors, packers and shippers, and other direct personnel who add value.

1.7.2 What? Quality Policy, Job Skills, and Audit Survival Training

Training for this group of employees is tailored to their job requirements and typically consists of the following aspects.

- **Quality Policy**

  This is usually conducted by the human resources department during new employee orientation and includes a briefing on the "Company Quality Policy" and procedures.

- **Job Skills Training**

  This is training every employee receives to do the job and usually is in the form of on-the-job-training (OJT) or internal classroom classes using documented work instructions.

- **Audit Survival Training**

  This training is usually conducted to prepare the front-line workforce for the registration audit. Table 5.9 provides an audit survival checklist which can be effectively utilized to conduct this type of training.

1.7.3 Why? To Implement Documentation

This training is required for workers to be able to perform their job correctly or, in other words, "Do what you say" as indicated by procedures and work instructions.

1.7.4 Where? Internally

This training will be performed as OJT or classroom in-house training.

1.7.5 When? Ongoing

This type of training is ongoing and begins as soon as documented procedures or work instructions are available.

1.7.6 Cost?

This cost is calculated per the internal training program.

Table 5.9: Audit Survival Checklist Example

Requirement	Example
• Management Responsibility	What is the company quality policy and what does it mean?
• Quality System	Do you have procedures that provide guidance to carry out your job?
• Document and Data Control	Are appropriate documents available and of the latest revision?
• Product Identification and Traceability	Are lot cards being utilized?
• Process Control	Are you monitoring process parameters and product characteristics?
• Inspection and Testing	Are you inspecting product per the procedures/instructions?
• Inspection, Measuring, and Test Equipment	Is all inspection equipment in your work area calibrated?
• Inspection and Test Status	Is product status clearly identified?
• Control of non-conforming Product	How do you identify non-conforming product?
• Corrective Action	What do you do when product is non-conforming?
• Handling, Storage, Packaging, Preservation, and Delivery	Are you handling (finger cots, etc.) and packaging product per the instructions?
• Control of Quality Record	Are all data sheets being filled out correctly and completely?
• Training	What training is required to perform your job and have you been trained?
• Statistical Techniques	Are control charts being filled out correctly and what do you do when out-of-control patterns appear?

1.8 Conclusion

This section provided detailed training requirements for a QS-9000 registration effort. The training requirements outlined will not only help achieve registration efficiently, but also help design, develop, and implement an effective quality system.

We cannot stress the benefits of training enough. Training ensures:

- A good understanding of the project so that a sound registration plan with a high probability of success can be developed.
- That all will speak the same language, that of QS-9000.
- The development of highly skilled employees, who are the most important resources of an organization.

Readers will soon realize that training costs represent a sound financial investment and will be repaid manyfold during the life of registration.

2 Audit

In the QS-9000 registration project, audit is part of the **A**nalysis in TAP and **S**tudy in PDSA.

ISO 8402 defines audit as "a systematic and independent examination to determine whether quality activities and related results comply with planned arrangements and whether these arrangements are implemented effectively and are suitable to achieve objectives."

2.1 Audit Classification

Audits are of several kinds:

- First party
- Second party
- Third party

Audit techniques are primarily the same, the differences arise from the scope and objectives and from who performs the audits.

2.1.1 First Party Audit

A first party audit is an internal audit of the quality system. It is performed by trained internal auditors and the objective is to measure quality system compliance to the QS-9000 quality system requirements and identify areas for continuous improvement.

2.1.2 Second Party Audit

A second party audit is an audit of the vendor or sub-contractor. The audit is conducted by the customer of the vendor and the objective is to determine if the vendor is able to meet established quality requirements.

2.1.3 Third Party Audit

A third party audit is a quality system registration audit. It is an audit of the company's quality system by an independent organization such as a qualified registrar who determines whether the quality system complies to the applicable requirements or not.

A QS-9000 registration effort involves a variety of first and third party audits as shown in Table 5.10.

What What explains the type of audit.

Why Why states the purpose of audits.

Who Who identifies the individual who will conduct the audit.

Where Where addresses the locations and scope of audits.

When When indicates the specific timing of audits.

Cost Cost gives approximation in USD for budget purposes.

Table 5.10: QS-9000 Project Audit Matrix

What	Why	Who	Where	When	Cost
Needs Assessment Audit (1st Party)	Provides clear picture of the existing quality system	Internal Audit Team/ Project Team	All applicable areas	After developing Internal Auditors	Internal Costs
Desk Audit					
Internal (1st Party)	To ensure quality system documentation meets QS-9000 requirements	Internal Audit Team	On-site	Upon completion of documentation	Internal Costs
External (3rd Party)	To determine if the company is ready for registration based on the text in Quality Manual	Registrar (Lead Auditor)	Registrar site	Upon completion of documentation	$1500
Mock Audit/ Pre-Registration Audit (1st or 3rd Party)	To identify and document system deficiencies & prepare for Registration Audit	Internal or External Auditors	On-site	Upon completion of Documentation & Implementation	Internal Costs/ $4000
Registration Audit (3rd Party)	To obtain registration	Registrar (Audit Team)	On-site	When Mock Audit results are satisfactory	$5000
Surveillance Audit (3rd Party)	To sustain registration	Registrar (Lead Auditor)	On-site	Every 6 months	$1500

2.2 Needs Assessment Audit (NAA)

The needs assessment audit is covered in Chapter 4, "Step on the Gas."

2.3 Desk Audit (External and Internal)

2.3.1 What Is a Desk Audit?

A desk audit is a review of the quality manual against the QS-9000 quality system requirements.

A desk audit may be of two types:

- Internal
- External

Internal Desk Audit

This is the final examination of the quality system documentation before it is submitted to the registrar for the external desk audit. This audit looks at all levels of the quality system documentation for compliance to the QS-9000 quality system requirements.

External Desk Audit

The is a review of the company's quality manual for compliance to the QS-9000 quality system requirements by a registrar.

2.3.2 Why Is It Performed?

A key objective of the internal desk audit is to determine if quality system documentation accurately reflects the overall policies and directives of the company before submitting it to a registrar. It must address all elements of the QS-9000 requirements and point to the related procedures.

The objective of the external desk audit is to evaluate the company's policies and determine if the company is ready for a registration audit.

2.3.3 Who Performs the Desk Audit?

The internal desk audit is performed by the internal audit team.

The external desk audit is performed by the registrar. The registrar's lead auditor will usually be selected to perform the desk audit as it contains the information needed to prepare the registration audit plan. Often, this is the first opportunity to interface with the external auditors and learn about their audit methods.

2.3.4 Where and How Is the Desk Audit Performed?

The internal desk audit is performed at the company site. The external desk audit is usually performed at the registrar site. The steps shown below explain the external desk audit process.

1. Company submits quality policy manual to the registrar.

2. Registrar reviews how well the quality policy manual addresses the applicable QS-9000 quality system requirements.

3. Registrar provides the company with a formal desk audit report (Figure 5.1) identifying potential deficiencies or "issues" that need to be discussed and/or addressed by the company.

Note: If, in the lead auditor's judgment, the company has the ability to address the issues elsewhere in the described quality system, he/she normally recommends to "continue the registration process."

4. Company closes out open issues by providing explanations or additional documentation.

Note: Closure of issues may require an on-site visit by the registrar.

Desk Audit Report	
Date:	10/25/9_
Company:	Carewell Mediproducts
Location:	Ramnager Road, Naintital
Standard:	QS-9000
Scope:	Manufacture of Disposable Syringes
Documentation:	Quality Policy Manual
Lead Auditor:	Gurmeet Naroola

The following desk audit is a detailed appraisal of the quality policy manual. This review was conducted to determine if any significant omissions or deviations from the quality system requirements exist prior to a formal assessment.

Summary:

Carewell Mediproducts has stated in their application that they address all 23 elements of the QS-9000 requirement. After reviewing the quality manual, the following elements may not be applicable:

4.7 –Purchaser Supplied Product

4.19 –Service

Several other elements of QS-9000 appear to have significant weaknesses as explained in the following review:

Figure 5.1 Sample Desk Audit Report (Page 1 of 2)

Detailed Review:

4.1 Management Responsibility

4.1.1 Quality Policy

Carewell Mediproducts' mission statement is signed by the Plant Manager:

"We will strive to deliver error-free, competitive products and services on time to meet and exceed our customer expectations."

ISSUES: None

4.1.2 Management Representative

Various organizations (marketing, purchasing, operations, etc.) and their responsibilities are described. There is no mention of the management representative.

ISSUES:

Who is the appointed management representative?
What are his responsibilities?

4.1.3 Management Review

Formal review of the entire quality system is not addressed in the manual.

ISSUES:

What reviews are made of the system?
Are all elements reviewed?
What are the inputs?
What are the outcomes?
What records are maintained?

4.1.4 Business Plan

A business plan is eluded to, but not described in detail.

ISSUES:

Does Carewell formally document and review business plans on a regular basis?

Figure 5.1 Sample Desk Audit Report (Page 2 of 2)

2.3.5 When Is a Desk Audit Performed?

The internal audit team should be conducting document reviews throughout the development of the quality system documentation. Once the internal audit team determines that they are comfortable with the results, it should be scheduled with and submitted to the registrar.

2.3.6 What Is the Cost of an External Desk Audit?

A typical desk audit review and report normally requires one auditor day and will run about $1500.

2.4 Pre-Registration (Mock) Audit

2.4.1 What Is a Pre-Registration Audit?

A pre-registration audit, also known as a "mock audit" or a "baseline audit," is a dress rehearsal of the registration audit.

2.4.2 Why Is It Performed?

It seeks to measure in detail whether or not the quality system that has been implemented conforms to the QS-9000 quality system requirements. It is of prime benefit to the company going through the registration process as it gauges readiness for the third party registration audit.

The registration audit is a costly endeavor. Management needs to establish a level of confidence that the registration audit will result in a recommendation for registration. A pre-registration audit provides a level of confidence.

2.4.3 Who Performs the Pre-Registration Audit?

This audit can be conducted by either the internal audit team or the registrar audit team. Both approaches have their benefits.

- **Internal Audit Team**

The internal audit team can do an extremely comprehensive audit and provide a final opportunity to fix any problems that might have been overlooked. Internal auditors know the system and can often zero in on subtle discrepancies.

- **External Auditors**

External auditors provide a fresh and unbiased evaluation of the quality system by having the opportunity to systematically locate deficiencies within an unfamiliar system (forest through the trees syndrome). External auditors often have the perspective to summarize and highlight significant areas of concern. Areas which may seem on the surface to have marginal or unimportant performance may be areas of concern in the quality system. External auditors will tend to focus on the big picture, relationship of the procedures to the requirements, and QS-9000 specific deficiencies.

Another benefit of an external mock audit is the chance for internal personnel to go through the process of being interviewed prior to the final registration audit.

Utilizing external audit resources helps to better prepare for the final registration audit.

Obviously, the costs involved are significantly higher. The cost is that of the registration audit, which equates to the price of auditor days, TLC, and report costs.

2.4.4 When Is the Pre-Registration Audit Performed?

This audit is conducted once the quality system has been properly documented and implemented. Timing is always the prerogative of the company being registered.

2.4.5 Where Is the Pre-registration Audit Performed?

The audit takes place at the company location(s). It includes all areas that have been described in the scope of registration. In other words, all areas of the quality system and all personnel involved are subject to the audit.

2.4.6 How Is the Pre-registration Audit Performed?

Internally, the mock audit should be conducted in the same fashion as the registrar would conduct the registration audit. Talk to the registrar and request a sample registration plan. This approach will help the company better prepare for the registration audit.

Note: If the pre-registration audit shows significant problems or non-conformances, it may be a good idea to re-evaluate final registration dates with the registrar.

Externally, the registrar should perform the audit identically to a registration audit. The audit plans, interviews, and documentation of findings should be the same. The only difference should be that there is no official outcome or recommendation. The benefit of "no outcome or recommendation" is that the findings are for the company's benefit. The non-conformances should be addressed prior to the final registration audit. Another benefit of the external audit is the chance for internal personnel to experience being interviewed, practice responding, and participating in an audit.

2.5 Registration Audit

2.5.1 What Is a Registration Audit?

The registration audit is a third party audit which verifies that the company's quality system is documented per the QS-9000 quality system requirements and assesses implementation of the procedures necessary to meet the quality standard. There must be objective evidence to support implementation of all elements and areas.

2.5.2 Why Is It Performed?

The registration audit is a necessary condition to achieve registration.

2.5.3 Who Performs the Registration Audit?

This audit is conducted by the third party registrar's audit team. The team will consist of all QS-9000 auditors and at least one automotive expert. A lead assessor will be a primary contact and provide schedules and other information related to the final audit.

2.5.4 Where Is the Registration Audit Performed?

The audit should be performed at all locations, sites, and work areas covered by the scope of the quality system to be registered.

2.5.5 When Is the Registration Audit Performed?

When the pre-registration audit results are satisfactory and the project team has confidence in the quality system. The team must evaluate all available information and schedule a reasonable date with the registrar. This date selected should give adequate time to implement any known non-conformances and establish history of consistent implementation.

2.5.6 How Is the Registration Audit Performed?

The actual audit comes after a great deal of preparation. First, the date and number of auditor days are established. Then, the registrar's lead auditor and the company's management representative communicate with each other to establish the schedule for the registration audit. Figure 5.2 shows a typical three-day/two-auditor audit schedule.

Carewell Mediproducts
QS-9000 Registration Assessment Schedule
June 22, 23, and 24, 199_
June 22, 199_

Times:	Dept. or Location:	Activity:	QS-9000	Assessor Number: 1	Assessor Number: 2
8:00 am	Conf. Room	Opening Meeting		X	X
8:30 am	Conf. Room	Management Responsibility	4.1	X	X
10:00 am	Conf. Room	Overview of the Quality System	4.2	X	X
		Quality Planning	4.2	X	X
		Manufacturing Cap.	Sect. II	X	X
11:00 am	Marketing	Contract Review	4.3	X	
		Customer Specific Requirements	Sect. III	X	
11:00 am	Purchasing	Purchasing	4.6		X
	Purchasing	Purchaser Supplied Product	4.7		X
12:30 pm		Break			
1:30 pm	Engineering	Design Control	4.4	X	
		PPAP	Sect. II	X	
		Manufacturing Capabilities	Sect. II	X	
		Customer Specific Requirements	Sect. III	X	
1:30 pm	Documentation	Document and Data Control	4.5		X
	Engineering	Document and Data Control	4.5		X
3:30 pm	Production Ctr.	Production Control	4.9		X
4:30 pm	Conf. Room	Closing Meeting		X	X

Figure 5.2 Registration Audit Schedule (Page 1 of 3)

June 23, 199_

Times:	Dept. or Location:	Activity:	QS-9000	Assessor Number: 1	2
8:00 am	Receiving & Incoming Inspection	Product ID and Traceability	4.8	X	X
		Process Control	4.9	X	X
		Inspection and Testing	4.10	X	X
		Inspection, Meas., and Test Equip.	4.11	X	X
		Inspection and Test Status	4.12	X	X
	▲	Control of Non-conforming Product	4.13	X	X
	Production	Corrective and Preventive Action	4.14	X	X
	▼	Handling, Storage, Pkg., and Del.	4.15	X	X
		Control of Quality Records	4.16	X	X
		Statistical Techniques	4.20	X	X
12:30 pm	Break				
1:30 pm	Continue above[a]	Same elements as above, through final inspection & shipping		X	X
				X	X
				X	X
				X	X
4:30 pm	Conf. Room	Closing Meeting		X	X

[a]The team will split up to cover different production lines.

Figure 5.2 Registration Audit Schedule (Page 2 of 3)

June 24, 199_

Times:	Dept. or Location:	Activity:	QS-9000	Assessor Number: 1	Assessor Number: 2
8:00 am	Calibration/	Inspection, Meas., and Test Equip.	4.11	X	
9:00 am	Tooling	Manufacturing Capabilities	Sect. II	X	
8:00 am	Human Resources Quality	Training	4.18		X
10:00 am		Corrective & Preventive Action	4.14	X	
11:00 am		Continuous Improvement	Sect. II	X	
10:00 am	Audit Team	Internal Quality Audits	4.17		X
12:00 pm		Break			
1:00 pm	Quality	Control of Quality Records	4.16	X	
2:00 pm	Service Dept.	Servicing	4.19	X	
1:00 pm	Quality	Statistical Techniques	4.20		X
3:30 pm	Conf. Room	Close Open Issues Team Caucus		X	X
4:30 pm	Conf. Room	Closing Meeting		X	X

Figure 5.2 Registration Audit Schedule (Page 3 of 3)

From the auditees' perspective, the project team should plan their participation in the registration audit in great detail. After receipt of the schedule from the registrar, the key personal should be identified and prepared for each element. Also, all who come in contact with the auditor should be prepared by making them familiar with typical questions that may be asked and the method of response. The better the preparation, the more efficient and effective the audit.

Preparation should include access to copies of relevant procedures, organization charts, master lists, and maps of the facility. Arrangements should be made for special equipment, eyewear, and protection for auditors.

The company can duplicate the schedule and elaborate on conference room names, contact names/numbers, shop locations, etc. Publish this company-wide. This helps both the auditee and the auditors.

Opening Meeting

The audit team arrives on site. Introductions are made. The lead auditor chairs the opening meeting and states the registrar responsibilities to the company, confirms the scope of audit, presents the agenda, and explains in detail how the audit will be conducted. Also, at this meeting the management representative identifies the escorts who will accompany the individual auditors.

Registration Audit

After the opening meeting, the audit team proceeds with the audit. Requirements such as management responsibility and quality system are typically audited first. The reason for auditing these elements first is that a major deficiency in either one may be sufficient grounds for terminating the audit. They also set the stage, so to speak, for the audit flow. The participation of top management reflects on the commitment of the project.

The auditors are very experienced in what they do. They are usually absorbing and evaluating many elements at one time, even without the escort or auditee knowing it. Just because there are not specific questions asked does not mean the auditor has not ascertained compliance with a requirement. For example, an auditor may notice that all shelves in a storeroom are all very clearly marked with product status. This may be enough evidence to avoid an additional question.

The moral of this story is, during the audit, do not continually worry about what is "not" asked. One should, however, be concerned about assumptions or findings that resulted in the absence of a question. It is the auditor's responsibility to dig into an issue to make note of a non-conformance. They must have objective evidence to make that judgement or determination and document a non-conformance. It may be the auditees' responsibility to dispel it.

Discussions about activities are directed to the individuals actually performing the work. The auditors will quickly pull an escort aside for a chat if the escort enjoys jumping in and answering all of the questions! Answers are expected from everyone other than the escorts unless there is a language barrier. The escort is also expected to step in when there are situations of extreme nervousness. In general, the auditor determines where to go, what to see, and who to talk to.

While auditing, the auditors take notes, both of non-conformities and conformity. Excessive writing on the part of the auditor should not be construed as "bad." For those instances where there appears to be a non-conformance, it is documented using some type of an observation form (Figure 5.3). For the observation to become a true non-conformance, the auditor must cite the applicable clause from the standard. "Observations," "Non-conformances," and "Findings" at a minimum state the company, date, area, escort, detailed finding, element of violation, and auditor.

Observation Form

Company Name: ____________________

Location: ____________________

Standard: ____________________

Date: ____________________

Location of Observation:	Escort:	Observation Number:

Observation:

Auditor:	Acknowledged by:

Requirement: (element# / brief description)

Major / Minor	Documentation ☐ Implementation ☐	CAR #

Figure 5.3 Observation Form

All issues identified during the desk audit that were not closed are also investigated and verified. If they remain open, they become observations and findings.

All areas are audited progressively until the audit is complete. It is common for auditors to revisit areas as they determine the need for additional information.

Daily Meeting

At the end of each audit day, the lead auditor reviews with other audit team members the status of the audit to ensure that it is progressing in a timely and efficient manner. The team will then invite representatives of the company to join the meeting. It is at these meetings that the lead auditor communicates any non-conformances to interested personnel and, at a minimum, to the management representative. Non-conformances are not usually identified as major/minor at this time. The registrar encourages discussion about the content of the findings. If other information can be brought to the table to counter a finding, the daily meetings are the appropriate forum.

The auditors do not have motivation to write unsubstantiated findings. Good auditors will readily get to the root of a non-conformance before even beginning to write. Nonetheless, there are times when an interviewee may get flustered or there is an oversight in what was represented. The auditors should be open to any and all discussion.

If a non-conformance turns out to have evidence that voids it, it may be reduced, removed, or turn into a different finding altogether.

Closing Meeting

At the conclusion of the audit, the lead auditor conducts an exit meeting with the auditees and presents a summary of all observations or findings. At this point, the observations will be formal non-conformances. The recommendation concerning registration is announced. Also, at this meeting the management representative acknowledges the results of the audit.

Audit Report

The lead auditor sends a registration audit report containing a summary of the audit and copies of all findings to the company's management representative, normally within ten working days of the conclusion of the audit.

The typical report will include the following:

- Report summary - summary of company, location, scope, standard, date of events, and auditor names. Summary will also have number of observations and recommendation/conclusion.
- Audit plan/matrix - listing of elements and areas covered. This normally has a cross reference of elements reviewed in each area/department.
- Agenda - audit plan with times and locations specified
- Interview list - list of names of personnel interviewed
- Opportunities for improvement - list of opportunities for improvement as identified by the auditors during the audit.
- Non-conformances - copies of the non-conformances to address

Other attachments will be registrar dependent and may include attendance lists, desk audit closures, etc.

Corrective Action

If a corrective action is required, the auditee provides a written corrective action plan to the lead auditor within a specified time period.

2.5.7 Cost?

A typical two-auditor/three-day audit will run up to about $9000.

2.6 Surveillance Audit

2.6.1 What Is a Surveillance Audit?

A surveillance audit is a scaled-down version of the registration audit. Typically it is a one-day audit conducted by the lead auditor.

2.6.2 Why Is It Performed?

It is a requirement, once the company has been registered, to ensure that the registered quality system remains in compliance to the QS-9000 quality system requirements.

2.6.3 Who Performs It?

The surveillance audit is usually conducted by one of the auditors who participated in the initial registration audit.

2.6.4 Where Is It Performed?

Similar to the registration audit, this audit is conducted on site.

2.6.5 When Is It Performed?

Per the QS-9000 requirements, surveillance audits are conducted every six months. The audits are not surprise visits and will be scheduled well in advance.

2.6.6 How Is It Performed?

The surveillance audit is conducted in the same manner as the registration audit. Figure 5.4 shows a typical surveillance audit plan.

Time	Activity
8:00 am	Review of current status of quality manual
8:15 am	Current use of the symbols
8:30 am	Closure and verification of open non-conformances (0:15 minutes per finding)
9:00 am	Re-evaluate element 4.2 (Quality System)
10:00 am	Re-evaluate element 4.6 (Purchasing)
11:00 am	Re-evaluate element 4.18 (Training)
12:00 pm	Break
1:00 pm	Re-evaluate element 4.1 (Management Responsibility)
2:00 pm	Re-evaluate element 4.14 (Corrective and Preventive action)
3:00 pm	Re-evaluate element 4.17 (Internal Quality Audits)
4:00 pm	Closing Meeting

Figure 5.4 Surveillance Audit Plan

- **Quality Manual Review**

  A typical surveillance audit visit begins with a review of the company's quality manual, with special attention to any major changes that have occurred during the last six month period.

- **Use of Symbols**

  Next, the QS-9000 symbols are evaluated to confirm that they are being used properly.

- **Verification of Non-Conformances**

  The auditor then verifies that corrective action for all recent non-conformances are satisfactorily implemented.

- **Audit**

  The auditor audits the quality system elements per the audit schedule. Besides the specific elements selected, management responsibility, corrective and preventive action, and internal quality audit are usually covered during each audit. The objective is to cover all quality system elements within a three-year period.

- **Closing Meeting**

  The closing meeting is held after the audit has been completed and the auditor summarizes his findings.

2.6.7 How Much Does It Cost?

A one-day/one-auditor surveillance audit will cost about $1500.

2.7 Conclusion

This section provided the reader with the necessary information to effectively prepare for and conduct the several audits that occur during the QS-9000 registration project. Shown are examples of actual audit plans, forms, and reports that provide as realistic a picture as possible of actual events.

In closing, the author would like to re-emphasize that audit is part of the **A**nalysis in TAP and **S**tudy in PDSA and is one of the more valuable continuous improvements tools.

3 Planning

Good planning will result in a successful registration project and is the basis of the TAP-PDSA approach. Planning should be as detailed as possible. Remember, "The devil is in the detail." Figure 5.5 represents the overall QS-9000 registration plan using the TAP-PDSA approach.

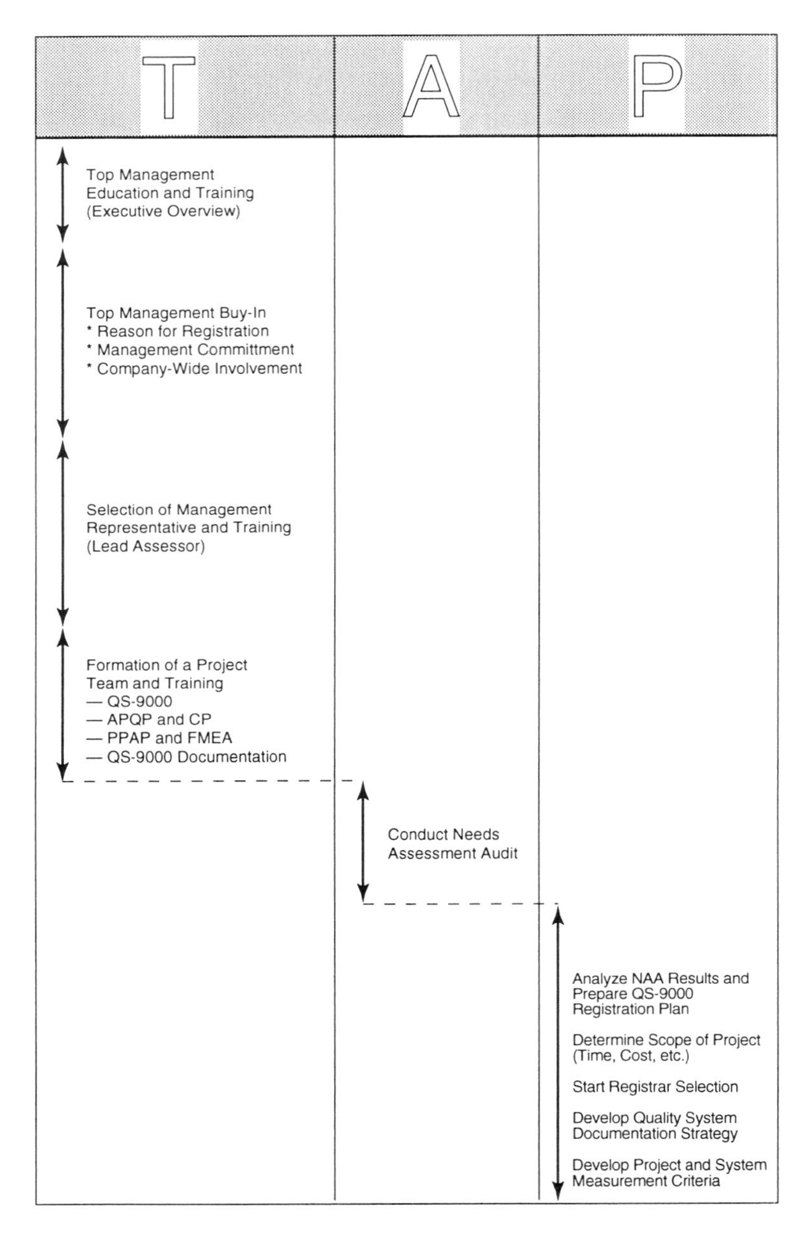

Figure 5.5 QS-9000 TAP PDSA Project Plan (Page 1 of 2)

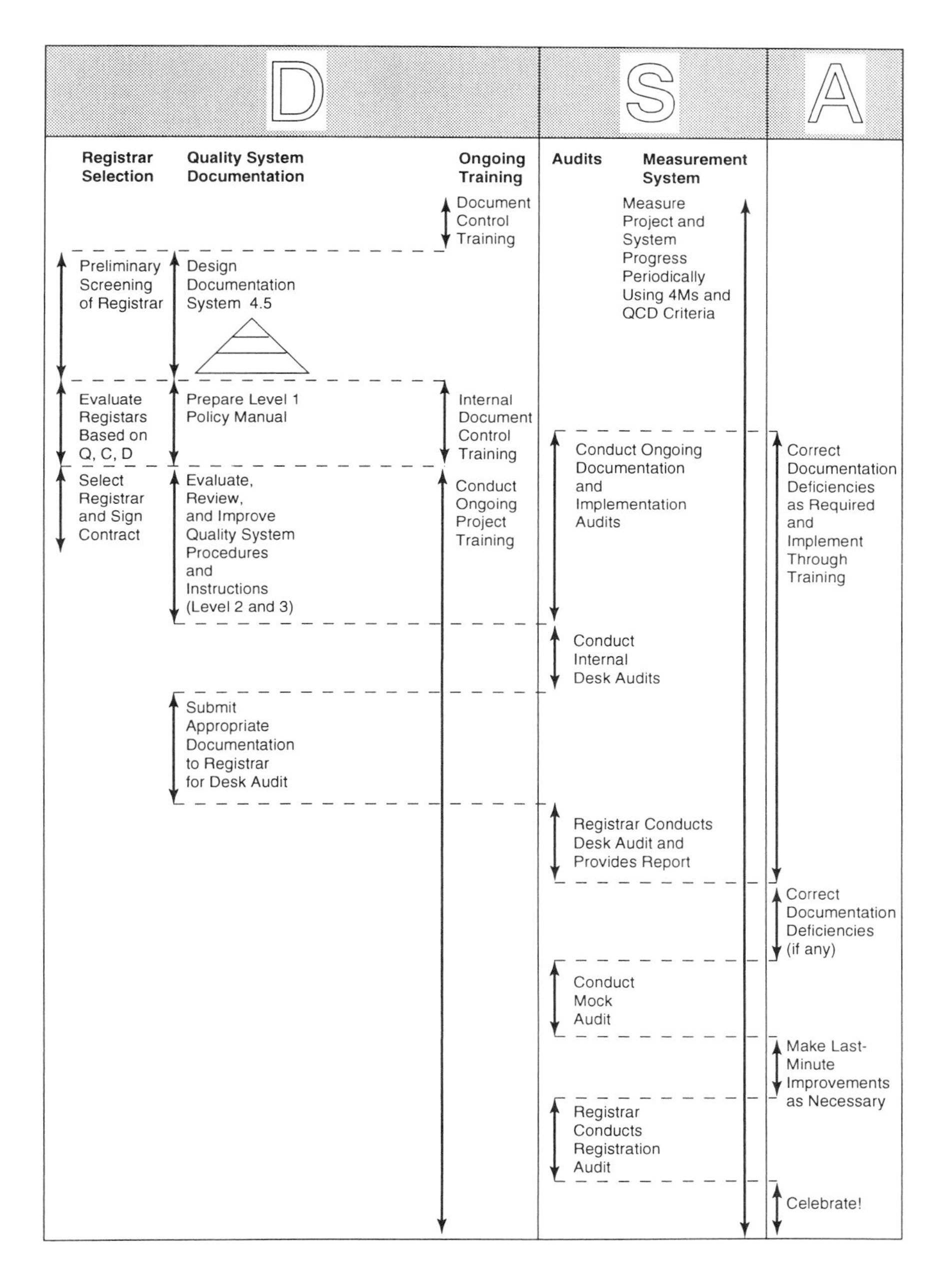

Figure 5.5 QS-9000 TAP-PDSA Project Plan (Page 2 of 2)

3.1 Training and Education

Phase one of the registration is training and education and involves getting the top management, management representative, and project team trained and educated in the QS-9000 registration effort. See Section 1 of this chapter for details regarding training.

3.2 Analysis

Phase two is the needs assessment audit (NAA), which provides a status quo/clear picture of the existing system. NAA is explained in detail in Chapter 4, "Step on the Gas."

3.3 Plan

Using the NAA information, the company develops the QS-9000 registration implementation plan for the key activities (measurement system, registrar selection, and QSD) explained in Chapter 4. Further information regarding measurement system, registrar selection, and QSD is provided in Chapters 6, 7, and 8, respectively.

3.4 Do

The plan is then implemented through the "do" step. This step represents the implementation of the plan.

3.5 Study

The project progress is checked through ongoing audits (internal and external) and other project measurements criteria. See Section 2 in this chapter regarding details on audits and Chapter 6 for further information on measurements.

3.6 Act

Any changes in the project are conducted through ongoing corrective actions and implementation, which represent the "act" stage.

The two cycles (TAP and PDSA) continuously evolve the quality system that results in an efficient QS-9000 registration.

3.7 Conclusion

Good planning makes the difference!

Remember, a plan is only a prediction with a probability of success, the probability of success being directly proportional to the amount and quality of training invested in the project.

Train and analyze the status quo before making a plan. Surely the project will then be successful.

Chapter 6
Driving Between the Lines (Measurements)

As stated in Chapter 4, measurements consist of the following:

- QS-9000 registration project measurements
- Quality System performance measurements

1 QS-9000 Registration Project Measurements

The registration must be completed on time, within budget, and should achieve the goal of successful registration.

During the registration project, measurements are often overlooked and not given the attention they deserve. For example, when people are asked about their registration project success stories, the answers are too familiar: "we improved our quality system," "communication increased," and so on. These answers are too general to represent any measure of success. Success must be quantified in "hard numbers" to be meaningful. Consider the answers as follows: system reject ratio decreased by 18%, documentation reduced by 21%, and customer audits reduced from 18 to zero. Now *that* is measurable success.

To obtain these measurements, the project team should develop a set of project measurement criteria. These are best developed using a matrix built around the 4 Ms and QCD (Figure 6.1).

The matrix consists of twelve interrelated cells, covering the entire quality system. Each cell indicates a possible measurement that can be used to track the project progress. However, the QS-9000 registration project is man and method oriented, and that is where most of the measurements criteria will appear.

For example, the man-cost cell contains training costs involved with the QS-9000 registration project. The training cost budget is estimated using the 5 Ws and 1 H training matrix as explained in Chapter 5. These costs are monitored by periodic measurement of the actual plan resulting in appropriate corrective action or modification.

Note: The matrix shown is generic. Develop a matrix of suitable measurements that meet the needs of your organization.

2 Quality System Measurements

The end objective of the quality system is to produce an output with an acceptable combination of quality, cost, and delivery that will result in customer satisfaction and profitability.

The QS-9000 requirement identifies these measurement issues by requiring the supplier's management to review the quality system at defined intervals, ensuring that goals and objectives are met. To emphasize this requirement two sections have been added to the management responsibility Element 4.1 in QS-9000.

- Analysis and Use of Company-Level Data
- Customer Satisfaction

	Man	Machine	Material	Method
Quality	• QS-9000 Training • Auditing Hours • Project Planning	• Number of Uncalibrated Equipment • Number of Process Changes	• Vendor Qualification	• No. of Procedures • Documentation Compliance Level • No. of Non-Conformances
Cost	• QS-9000 Training Costs (Internal and External) • Auditing Costs • Planning Costs • Registrar Costs • Consultant Costs	• Calibration Costs • Process Change Costs	• Vendor Qualification Costs	• System Development Costs (Management Responsibility, Quality System, . . . Statistical Methods)
Delivery	• Project Schedule (Activity Against Time) • Training Completion • Auditing Completion	• Calibration Compilation • Process Change Completion	• Approved Vendor List (AVL)	• System Completion (Documentation and Implementation Compliance Levels)

Project Objective:

- Successful QS-9000 Registration

Figure 6.1 QS-9000 Project Measurement Matrix

2.1 Analysis and Use of Company-Level Data

According to QS-9000: "The supplier shall document trends in quality, operational performance (productivity, efficiency, effectiveness) and current quality levels for key product and service features.

Trends in data and information should be compared with progress towards overall business objectives and translated into actionable information."

2.2 Customer Satisfaction

QS-9000 also states: "The supplier shall have a documented process for determining customer satisfaction, including frequency of determination, and how objectivity and validity are assured. Trends in customer satisfaction and key indicators of customer dissatisfaction shall be documented and supported by objective information."

The bottom line is that the system must be continuously measured to determine the following:

- Are the customers happy?
- Is the company making money?

Shown in Figure 6.2 is a quality system measurement matrix utilized to measure the above.

The system measurement matrix uses the same twelve-cell approach as the project measurement matrix. Each cell indicates a possible measurement that can be used to track the system output or the "health" of the system, which in turn relates to customer satisfaction and profitability.

Some companies use these techniques to develop customized quality indices depicting the overall health of the system.

	Man	Machine	Material	Method
Quality	• Employee Morale • Employee Turnover • Worker Output/ Productivity • Training Hours per Employee • Number of Suggestions per Employee • Number of Complaints • Absenteeism	• Machine Efficiency • Machine Capability (Reproducibility and Repeatability) • Machine Flexibility	• Reject/Scrap Rate (ppm) • Number of Non-Conformances (Internal and External) • Lots Defective • Customer Returns • Process Capability Indices (CPK) • Average Outgoing Quality Level (AQL)	• Organization Levels • Audit Non-Conformances • Number of Document Changes • Number of Engineering Changes • Corrective Action Effectiveness
Cost	• Sales per Employee • Recruiting Costs • Training Costs • Direct and Indirect Costs • Unplanned Overtime Costs • Absenteeism Costs	• Downtime Costs • Repair Costs • Spare Parts Costs • Equipment Costs • Maintenance Costs	• Scrap Costs • Rework Costs • Inventory Costs • Material Costs	• Document Control Costs • System Costs (MIS, MRB) • Activity-Based Costs (ABC)
Delivery	• Number of Meetings • Complaint Resolutions • Average Employment Duration	• Repair Time • Space Availability • Maintenance Response Time	• Just-In-Time (JIT) • On-Time Shipments • Number of New Customers and Suppliers • Material Shortages • Rework Time	• Corrective Action Completion Time • Document Release Time • Information Response Time

System Objectives:
- Customer Satisfaction
- Profitability

Figure 6.2 QS-9000 System Measurement Matrix

3 Conclusion

The value of measurements in managing the quality system cannot be overemphasized. Measurements are an integral part of management's effort for continuous improvement.

The measurement system should be continually evaluated, always determining if it is providing management with the desired key information to keep the business in control.

Chapter 7
Registrar Selection

Chapter 4 explained the importance of early registrar selection and the selection steps. This chapter covers in detail the registrar evaluation process using the quality, cost, and delivery criteria.

This evaluation is usually done in two steps:

- Preliminary Screening
- Final Screening

1 Preliminary Screening

Preliminary screening involves the examination of all registrars available to narrow the list down to the top few. The task is accomplished using the following questions:

Q1) Is the registrar approved to conduct QS-9000 audits?

There is a published list of QS-9000 approved registrars. This list is available to potential customers and includes those registrars who have met the criteria imposed by the Big Three and are officially approved. Only approved registrars may use "QS-9000" on issued certificates. If the registrar is on this list, they have met stringent criteria including:

- All audit members are QS-9000 qualified.
- Audit audit teams have an automotive expert per Big Three guidelines.
- Registrars passed a witness audit by the accreditation body.

Note: The list is available by contacting ASQC. If you have questions over and above the approved status, the registrar can assist you.

Q2) Does the registrar perform registration to the company's Standard Industry Code (SIC)?

In addition to QS-9000, a registrar should have qualification and expertise in registering companies in your SIC code. The registrar's SIC code qualifications are provided by formal documentation and often includes use of a technical expert hired for that express purpose. In any event, ask the registrar if they are qualified and ask them for documentation.

Q3) Does the registrar have acceptable credentials and references?

"By their reputation they shall be known." Call the management representatives of companies the registrar has registered. This is a common practice and most will be happy to accommodate as they went through the same effort. This is the most effective and expeditious method of learning about and screening registrar candidates. This important step is ignored by most.

Q4) Is there a conflict of interest?

ISO/IEC guide 48 states: "An organization that advises a company on how to set up its quality system or writes its quality documentation should not provide assessment services to that company unless strict separation is achieved to ensure that there is no conflict of interest."

Companies should do the necessary research to ensure that there is no conflict of interest, as this alone could jeopardize the company's registration effort. Registrars who also provide consulting services cannot perform registration within two years of consulting. In-house "training" is considered consulting. Since QS-9000 is relatively new, there is a waiver for registrars who provided training prior to August 1994. The waiver allows registrars to grant registration to a company provided consulting/training on behalf of that company ceased prior to August 1994.

Q5) Is the registrar available?

The registrar must be available for the planned registration audit dates. Demand for registrars is heavy, and one should not expect a registrar to be available on short notice. This is one of the key activities that we present in Chapter 4. Be sure that the registrar has audit resources available within your estimated time frame.

2 Final Screening

Once the top few registrars have been determined, it is time to select the registrar who best meets the company's needs.

Use the Quality, Cost, and Delivery (QCD) criteria to facilitate selecting the best registrar. Remember, anyone can achieve the best measure in two out of three. Get a balance of all three to obtain the best "value" for the registration dollar.

2.1 Quality Criteria

Q1) Does the registrar have a good quality history? (How long have they been around? How many companies have they registered? etc.)

Like any other supplier of a service, it is important to check the quality history of each registrar. Ask for the quality manual, and check the non-conforming and corrective action records. Also check with the accreditation body to find out if any complains have been filed against them. Rank the registrars by both the length of time they have been in business and the number of registrations they have done, and the absence of complaints.

Note: Many companies ask the registrar candidates to complete a supplier survey questionnaire.

Q2) What is the average caliber of the registrar's auditors? Are the auditors assessors or lead assessors?

The strength of any registrar is in the caliber of the audit staff. Ask for auditor backgrounds and check their experience based on number of QS-9000 and ISO-9000 audits plus years of audit experience. Ensure that the registrar staff consists of a high percentage of permanent auditors.

Q3) Will the registrar meet with the company for an interview prior to selection?

If a representative of a registrar is available locally, they will likely meet with the company. Often, the sales representative will have the most flexible schedules to enable meetings in person. The sales staff is a wealth of information, and one can learn a lot about the registrar by interacting with them. Many sales staff also provide audit support.

It may also be possible to interview an assessor who would likely be on the company's registration audit team. Due to the enormous amount of travel and busy schedules of assessors, they may not be able to accommodate the request. But it does not hurt to ask, particularly, if you are down to one or two registrars and you feel it is important. High scores should be given to registrars who attempt to meet the company's specific requests.

Q4) Does the registrar treat the company like a customer? How comfortable is the company with the registrar's personnel and methods?

Is the registrar user-friendly? Check very carefully for a reading on how registrars conduct business with their clients. A company will have a business relationship with the registrar for a number of years and it should be based upon trust, mutual respect, dependability, flexibility, and a genuine interest in serving the needs of the client.

Q5) Does the registrar understand the way the company conducts business?

The registrar must take time to understand the company's type of business especially in a fast-changing environment such as high-technology industries where products, processes, and methods are continually being changed and improved, sometimes in a matter of months.

Q6) How are registrar complaints handled?

The registration process is a two-way street and should provide for a third party problem resolution process. Open discussion should be a matter of practice from the start, and discrepancies can be discussed as they arise. There are also formal avenues for complaints during the registration process. The registrar should remember that the company is the customer and examine the problems from the company's point of view. High scores should be given to registrars who indicate they deal with any customer complaints in an efficient/professional manner

Q7) What happens if the registrar goes out of business?

Similar to a supplier selection, some level of confidence that the registrar will be around in future years is needed. This is done through the examination of the registrar's registration count and market share.

The future of any corporation, however, is not certain. It is important to understand the legal ramifications of the contract and have confidence that the registrar selected will honor the contract. High marks should go to registrars that have a stable base of registered companies and a respected reputation.

2.2 Cost Criteria

Q1) What is the registrar's total cost of registration (three years)? Does it meet the company's budget?

Cost of registrars varies considerably and involves significant expenditure of money. The two largest cost factors are labor costs (auditors and reports) and logistical costs (transportation, meals, lodging). While calculating the costs, consider the total registrar costs and not just the initial costs. There is a lot more than meets the eye.

Consider the following activities while determining the total registrar costs:

- Application fee
- Preparation and initial visit
- Desk audit and revision review
- Pre-audit
- Registration audit
- Certificate
- Surveillance audit (five additional in three-year cycle)
- Overall travel and living costs (TLC)

Note: Remember that the registrar is a supplier and cost negotiation is recommended. Although fee structures are carefully constructed, registrars may negotiate fees when appropriate to gain business.

Q2) Is the registrar cost structure fixed for the life of registration?

Negotiate a contract requiring firm cost for a period of three years.

Note: Registrar costs should decrease as the availability of registrars increases and business becomes more competitive.

Q3) How can a contract be canceled?

Learn the contract cancellation policies of the registrars. At some point an organization may decide to choose another registrar or give up registration for any number of business reasons. The costs associated with contract cancellations should be minimal.

2.3 Delivery Criteria

Q1) Is the registrar flexible regarding audit dates, etc.?

Registrars with extensive backlogs tend to provide limited opportunities for rescheduling. Even the best schedules sometimes require changes and the registrar should be able to accommodate without undue difficulty. In such a scenario, this criterion becomes an important factor. The ideal registrar is available on demand. Try finding a registrar who most closely approaches "availability on demand."

Most registrars will have fixed rules to be protected from last minute changes which leave their auditors idle. High marks should be given to registrars who have fair and reasonable change/cancellation policies. An example of fair may be a fee charged for cancellation of audit less than 30 days in advance.

Q2) Does the registrar have a local office and auditors?

Having a local registrar and auditors represents significant benefits:

- Auditor TLC (transportation/airfare/ hotel etc.) are lower.
- It is possible to establish a closer relationship with a local organization. It also gives the registrar an opportunity to learn more about the company's type of business and provide enhanced service.
- It provides an opportunity to meet and talk to the auditors and get to know their auditing attitudes and techniques.
- The local office representative can be a great networking source for providing training referrals and other information.

Q3) How does the registrar accommodate changes in scope or content of registration?

Changes in scope may require an amendment to the contract. Discuss the company's business direction and its effect on the scope of registration, especially if new sites and new products are added on. Some registrars may require a complete re-registration, while others may require only a change in the scope statement. High marks should be given to registrars who have a formal process for this.

Q4) What is the registrar's re-certification frequency?

Most registrars do not require re-auditing of the entire system after the three-year registration period. Certification holds good as long as the company continuously passes the surveillance audits and the surveillance audits cover the entire system in the course of three years. If the company does not desire a full re-certification audit after three years, select a registrar that does not require it.

Q5) How does the registrar audit against changes to the QS-9000 quality system requirement?

Each accreditation organization lays down the framework, and each registrar works within the framework. When a new version of QS-9000 or any other standard is issued, the effective date will be established. Usually a convenient surveillance date is selected as the implementation verification date. Some registrars may elect to dedicate a special one-day audit to review the changes. The bottom line is that the registrar must give their clients sufficient time to comply to the changes and the full implementation date must be mutually agreeable. Again high marks should be given to registrars who have formal processes to implement changes.

Q6) Can a company get the same auditors every time, and can they select auditors?

It is to the company's advantage to have audits performed by the same group of core auditors each time. If this is desirable, a good registrar should be open to this request.

Having "repeat" auditors leads to continuity. Less time will be spent establishing communications, understanding auditor audit methods, and explaining the complexities of the quality system. An auditor with a previous understanding of the quality system may conduct the audit more skillfully. Employees feel most comfortable having familiar faces visit each time. Keep in mind, this does not mean every auditor will have total recall of the processes as they deal with many other companies.

Q7) What is the registrar's response time to the desk audit and other reports?

Most registrars are good about meeting time schedules. A commitment on how fast the registrar will get audit reports back is necessary. Delayed reports, especially during the desk audit phase, can disrupt the company's registration timeline. Turnaround time for all reports should be built into the contract. Registrars usually state turnaround time in their quality manual.

Q8) How does the registrar accommodate consultants?

Many companies use consultants to help them in their registration efforts. A good question to ask the registrar is in what capacity the consultant will be allowed to participate during the registration audit. Many registrars disallow a consultant's presence; others will allow a consultant's presence without participation, and yet others may allow full participation. Consider registrars who accept consultants in the way the company wishes to use them.

3 Registrar Rating Sheet

Figure 7.1 shows a Registrar Rating Sheet developed using the QCD criteria to aid in the registrar selection process.

	Criteria	Registrars			
	Quality:	**A**	**B**	**C**	**D**
Q1)	Does the registrar have a good quality history? (How long have they been around? How many companies have they registered? etc.)				
Q2)	What is the caliber of the registrar's auditors? Are the auditors assessors or lead assessors?				
Q3)	Will the registrar meet with the company for an interview prior to selection?				
Q4)	Does the registrar treat the company like a customer? How comfortable is the company with the registrar's personnel and methods?				
Q5)	Does the registrar understand the way the company conducts business?				
Q6)	How are registrar complaints handled?				
Q7)	What happens if the registrar goes out of business?				
	Cost:				
Q1)	What is the registrar's total cost of registration (three years)? Does it meet the company's budget?				
Q2)	Is the registrar cost structure fixed for the life of registration?				
Q3)	How can a contract be canceled?				
	Delivery:				
Q1)	Is the registrar flexible regarding audit dates?				
Q2)	Does the registrar have a local office and local auditors?				
Q3)	How does the registrar accommodate changes in scope or content of registration?				
Q4)	What is the registrar's re-certification frequency?				
Q5)	How does the registrar audit against changes to the QS-9000 quality system requirement?				
Q6)	Can a company get the same auditors every time, and can they select auditors?				
Q7)	What is the registrar's response time to the desk audit and other reports?				
Q8)	How does the registrar accommodate consultants?				
	Section Total				

Figure 7.1 Registrar Rating Sheet

- Take the top few registrars based on the initial screening and rate them under the individual QCD criteria taking into account the explanations provided earlier in this chapter. Use any rating system (1,2,3 etc.). The important thing is to identify what the company considers as important criteria.
- Compute individual quality, cost, and delivery totals, and apply importance factor as appropriate.
- Compare and select registrar based on the best balance of QCD requirements.

4 Conclusion

Registrar selection is a crucial process in the registration effort, and a significant amount of energy should be devoted to it.

The rating system provided in this chapter was used for registrar selection by the author in several registration efforts. The list used to make up the rating system is by no means complete and it should be modified to meet the company's specific requirements.

Good luck in the registrar selection process!

Chapter 8
The Owner's Manual

A number of surveys have indicated, without exception, that most registration efforts fail because of problems with quality system documentation (QSD). Virtually all quality system standards and requirements are in agreement on one point: an organization must thoroughly document what it does. Good documentation is the heart and soul of any quality system.

Until recently, there were many manufacturing arenas where segments of the process were performed repetitively, by the same workers, for infinite lengths of time. The need for documentation was minimal. In the craftsmen era, the entire process was carried out by artisans. Methods were stored in the master's head and doled out to the apprentices on a need-to-know, rote instruction basis.

No longer!

In today's rapidly changing global business environment, systems and processes are changing continually, and it is impossible to remember anything but the most critical bits of information. Everyone involved in the quality system, from top management to front-line worker, manages with information and this information is in the form of documents. Envision a world without documents. Impossible!

The attribute that made the ISO 9000 series of standards the de facto standards used in the world today is creation of a format for consistency of quality. It is like a common language. This is possible only when the quality system is fully, carefully, and correctly documented. QS-9000 does not differ from this concept.

This chapter explains how to prepare a documented quality system consistent with the requirements of QS-9000 by addressing the following questions:

- Why do we need QSD?
- What are the QS-9000 QSD requirements?
- How is QSD organized?
- Who prepares QSD?
- How do we develop QSD?
- How do we implement QSD?

1 Why Do We Need Quality System Documentation?

Quality system documentation (QSD) is needed because it is a QS-9000 requirement. More importantly, QSD is a business necessity for a variety of reasons. The most important ones are:

- **Communication**

  QSD is a formal vehicle to communicate the company's quality management policies, procedures, and system.

- **Implementation**

  QSD provides specifics for implementation of quality-related activities. The "hows" of implementation.

- **Provides Control**

  QSG specifically states the criteria of acceptance in a concrete manner. When it is documented, there can be no confusion.

- **Provides Consistency**

  QSD provides consistency through changing environments, shifts, and personnel.

- **Trains Personnel**

  QSD assists in training individuals in how to perform their task. Good documentation is a valuable resource to use during on-the-job training.

- **External Purposes**

  QSD provides guidelines for conducting business using a common language. It facilitates customer and other external audits.

2 What Are the QS-9000 Quality System Documentation Requirements?

QS-9000 very specifically defines the quality system documentation requirements by stating:

"The supplier shall prepare documented procedures consistent with the requirements of the International Standard and the supplier's stated quality policy."

In QS-9000, there are a total of 20 ISO 9000-based elements and three specific Section II QS-9000 elements. All company processes, documentation, and records will likely fall into one of these 23 slots.

Table 8.1 through Table 8.23 show the 23 elements with a summary of the documentation and record requirements. This means, that these items must be documented in one of the levels of documentation, policy, or procedures. Typical records are also listed. The list is not exhaustive, and each company should research and compile a comprehensive list. These records should not be confused with procedures. Some records, however, are living, such as a "Business Plan." Business plans of previous years are records. As well as being an element in itself, these records are considered strong "objective evidence" that a company is following its procedures. A "—" indicates that no additional records are required.

Table 8.1: Management Responsibility

Required Documentation	Records
Quality policy	—
Quality objectives and/or Company level data	List of objectives plus records of periodic update
Interrelation of personnel	Organization chart
Responsibilities and authorities	—
Management representative	—
Management review process	Management review record (minutes) plus relevant supporting data
Business plan process	Current and prior business plans
Customer satisfaction process	Customer data and results

A management review procedure is not specifically required, but it certainly is recommended. The procedure should include the frequency, attendees (executive management), agenda items (what is reviewed including policies/procedures, objectives, corrective actions, internal audits), record requirement, and handling of action items as a result of the reviews.

Table 8.2: Quality System

Required Documentation	Records
Quality manual including a reference to related procedures	—
Quality system documents	—
Quality planning process	Quality plans, control plans, and FMEAs per APQP and FMEA references

There are numerous quality planning records as a result of the APQP and control plan processes, many of which are related to design. While documenting and implementing these processes, the specific records generated as a result of implementation should be identified and maintained. Some will likely be records related to Element 4.4, Design Control, while others like the final control plans may be linked to Element 4.2, Quality System. They should, however, be captured somewhere.

Table 8.3: Contract Review

Required Documentation	Records
Procedures on how a contract or order is processed including feasibility analysis and review	Quotes, orders, feasibility review records, customer-specific characteristic evaluations, and other documents related to negotiations
Process for amending contracts and communication of changes	Change orders and confirmations

This element requires a procedure for review, acceptance, and change of a customer order. For custom or new products, the contract review process becomes intertwined with the APQP process. Customer input and special characteristics are integral parts of the planning and design effort in APQP. The procedures should reflect the actual process.

Table 8.4: Design Control

Required Documentation	Records
Procedures for the overall design process including design/development planning	Milestone charts with tasks and responsibilities, project status reports, control plans
Qualification status of personnel	Training or competence records for CAD, simulation, GD&T
Organizational interfaces	Milestone charts, plans, reviews
Design inputs	Drawings, specs, design goals, customer waivers related to absence of CAD/CAE system
Design outputs	Drawings, specs, simulation results, flow charts, MSA plans, process procedures, test results, design FMEAs, studies to optimize, etc.
Design reviews	Records of review for each planned design review with agenda/attendees
Design verification and validation	Prototype acceptance, performance and reliability test results
Design changes	Customer approvals and waivers

The APQP and Control Plan manual plus the PPAP manual have detailed information on the various inputs and outputs of design phases. Only an exhaustive evaluation of the design process with that of QS-9000 will yield the correct procedures and records.

Table 8.5: Document and Data Control

Required Documentation	Records
Procedure for the control of all documents and data:	—
Process for approval and issue of all documents (internal and external)	Evidence of review and approval is usually on the document itself, may be on a change order
Identification of all controlled documents and establishment and availability of a master list	—
Process to review customer documents/specifications	Records of review with date of implementation
Process to change documents	Engineering change orders, document change orders

There are many documents and lists created for use in the control of documents and data. These documents, such as the master list, are not necessarily quality records that need to be maintained after their usefulness is realized and are actually controlled documents themselves. Engineering change orders (ECOs) are commonly maintained as a quality record, as they demonstrate conformance to requirements and provide valuable histories.

Table 8.6: Purchasing

Required Documentation	Records
Procedures to ensure purchased product meets specifications	
Use of customer-approved subcontractor list	The list is a controlled document, not a record
Selection, approval, and ongoing maintenance (control) of subcontractors	Records of evaluation and approval of suppliers, surveys/audits, ongoing supplier quality ratings/results
Subcontractor development process and scheduling to assist in 100% on time delivery	May include surveys/audits, planning activities to develop subcontractors
Processing of purchase orders	Purchase order review

Table 8.7: Control of Customer-Supplied Product

Required Documentation	Records
Procedure for storage and maintenance of customer owned product including tooling and returnable packaging	Records of reporting lost, damaged, or unsuitable product back to the customer

Table 8.8: Product Identification and Traceability

Required Documentation	Records
Procedures to ensure product is identified throughout the process	Product traveler, test reports and results, serial number documents

The records for identification and traceability are often the same as process control records or test status records.

Table 8.9: Process Control

Required Documentation	Records
Procedures to control production	Product traveler, router, order
Process to ensure compliance to government standards	Certificates and letters of compliance
Process to control customer special characteristics	Inspection and test results, evaluation results, SPC
Preventive maintenance program	Logs/results of preventive maintenance activities, predictive maintenance studies
Process monitoring and operator instructions	Product/process traveler, router, orders
Process capability requirements and ongoing performance tracking (PPAP)	Capability studies for each process, control charts, control plans
Process to verify job set-ups	Record of verification, logs
Process change procedure	Change effectivity date record, other records per PPAP
Verification of appearance items	Qualified personnel records

Table 8.10: Inspection and Testing

Required Documentation	Records
Procedures related to testing and inspection of product to ensure product conforms to specified requirements	
Receiving inspection and testing	Inspection/test records, logs, subcontractor history cards, sub-contractor statistical data, urgent production release records
In-process and final inspection and testing	Inspection/test records, logs, travelers, urgent release records for in-process inspection only
Layout inspection	Record of layout inspection and functional verification per customer specific requirements

The Standard specifically requires that the inspection authority be identified on inspection and testing records. In other words, the inspector or tester should sign, stamp, initial, or log their name/ID. Records can obviously be electronically maintained as is common in today's environment. In that case, the product may be electronically tracked in a database, and the inspection authority is usually provided for by the individual's login ID.

Table 8.11: Inspection, Measuring, and Test Equipment

Required Documentation	Records
Procedure to identify, calibrate, and maintain equipment related to the acceptance of product	Records of calibration, certificates of calibration from external calibration services

Table 8.11: Inspection, Measuring, and Test Equipment

Required Documentation	Records
Control of test software plan	Log or records of checking of test software
Process to evaluate out-of- tolerance situations	Record of evaluation of the extent of situation (corrective action)
Measuring system analysis (MSA reference manual)	Studies for each applicable piece of equipment

Depending on whether calibration is performed internally, externally, or both, inspection, measuring, and test equipment records will be of various types. The individual assigned the responsibility to review the calibration records will initiate the process to evaluate out-of-tolerance situations.

Table 8.12: Inspection and Test Status

Required Documentation	Records
Procedures to ensure product pass/fail test status is identified throughout the process	Product traveler, test reports and results, serial number documents

Table 8.13: Control of Non-Conforming Product

Required Documentation	Records
Procedure to identify, review, and disposition nonconforming product and suspect product	Records of non-conformances, records of dispositioning, record of rework
Process to control reworked product	Results of analysis of non-conformances and reduction plan

Table 8.13: Control of Non-Conforming Product (Continued)

Required Documentation	Records
Engineering approved product authorization	Records of customer approval with date and quantity limitations for product requiring such authorization

Nonconforming product is often identified and documented by using a standard form containing the appropriate routings for evaluation and disposition. A Material Review Board (MRB) is often employed.

Table 8.14: Corrective and Preventive Action

Required Documentation	Records
Procedure to analyze the causes of nonconforming product and take action to correct the non-conformance	Corrective action records with type of non-conformance, root cause analysis, action taken, preventive action, methods to ensure closure, demonstration of effectiveness
Process to evaluate returned customer product	Failure analysis reports, corrective action records
Process to identify and track data for sources of potential non-conformances to initiate preventive action	Trend data, analysis, and subsequent corrective action records

There is a distinction between corrective and preventive action when discussing the *source* of addressing the non-conformance. A corrective action will be generated as a result of an incident of non-conformance and is reactive in nature. A specific customer complaint is an example of what would turn into a corrective action.

A potential preventive action will be identified as a result of analyzing data and observing a trend. For example, the nonconformance in this case may be the worst offender of last month's rework data. The investigative process and documentation will likely be the same for both a *reactive* corrective action and a *proactive* preventive action.

Table 8.15: Handling, Storage, Packaging, Preservation, and Delivery

Required Documentation	Records
Procedure for handling, storage, packaging, preservation, and delivery	Inventory records
Process to monitor delivery performance	Data, reports, shipment notifications to customer in event of system shutdown

Table 8.16: Quality Records

Required Documentation	Records
Procedure to identify, collect, index, maintain, file, store, and dispose of records	—

The quality records documentation should include the methodology for all the requirements of the standard, including where (location of filing) and how (method of indexing), etc. The documentation should allow the reader the ability to readily locate any quality record. Records that are maintained electronically are maintained, indexed, stored, and disposed of differently than paper and procedures should reflect the details.

Table 8.17: Internal Quality Audits

Required Documentation	Records
Procedure to plan and conduct internal quality audits to determine if the system is effective and complies with planned arrangements	Audit schedule, element by area matrix
Stipulation of trained and independent auditors	Audit assignments, schedule
Process for conducting and recording the audit results and deficiencies	Audit checklist, summary reports with compliances and deficiencies (non-conformances)
Method for conducting follow-up activities	Follow-up audit reports, detailed closure of corrective actions as a result of non-conformances

It is important to note that the audit record should include areas of conformance, as well as non-conformance, to give evidence that the audit was conducted.

Table 8.18: Training

Required Documentation	Records
Procedure to identify training needs and provide for the training of all personnel affecting quality (everybody)	Training records including date, type of training, trainer, trainee, certifications
Method for evaluating the effectiveness of training	Reports, data analysis related to performance, testing, surveys

Table 8.19: Servicing

Required Documentation	Records
Procedure for after-sales servicing processes	All applicable records from the QS-9000 elements as identified in these tables

Table 8.20: Statistical Techniques

Required Documentation	Records
Method for evaluating the need and use of statistics	Control charts, logs, reports, results

Table 8.21: Production Part Approval Process (PPAP) (Section II)

Required Documentation	Records
Procedure implementing the PPAP process	Part submission warrants, appearance approval reports, plus other PPAP specific documents depending on the submission table

The PPAP submission has many variables and the records are specific to the circumstance. The PPAP process must be thoroughly implemented.

Table 8.22: Continuous Improvement (Section II)

Required Documentation	Records
Plans for the implementation of a comprehensive continuous improvement process	Specific action plans for processes important to the customer, results, listing of opportunities for future improvement

Table 8.23: Manufacturing Capabilities (Section II)

Required Documentation	Records
Procedure for tooling management	Records of tracking and tool changes

The 23 elements stated above go hand-in-hand as the outputs of most processes are products and residual records. The records require special control per Element 4.16 Control of Quality Records.

3 How Is Quality System Documentation Organized?

Most quality system documentation is organized in four levels as shown in Figure 8.1.

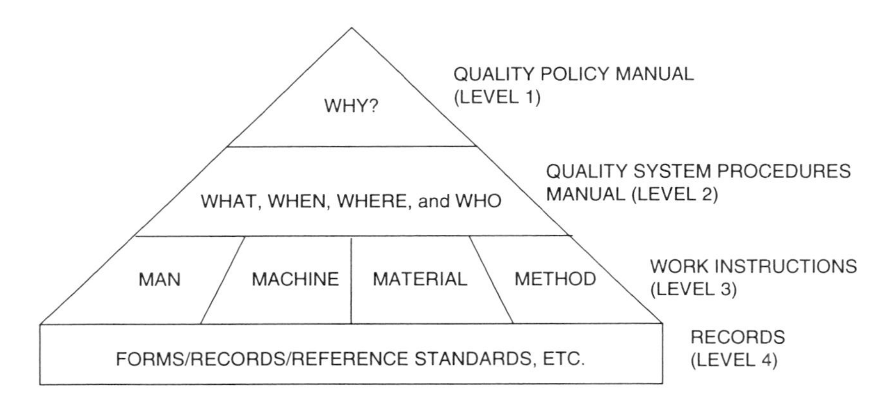

Figure 8.1 Quality System Documentation Organization

3.1 Quality Policy Manual/Level 1

The Quality Policy Manual (QPM) should describe the essence of the company's quality system. This manual defines, in policy, the quality system and addresses all applicable elements. Here the "Why are we doing all this?" is stated. The Quality Policy Manual is normally prepared by the Quality Assurance department. It is short and concise, does not contain any proprietary information, and is usually kept to one page per requirement. This is the manual that is sent to the registrar for the desk audit and is also available to new or existing customers.

3.2 Quality System Procedures Manual/Level 2

ISO 8402 (Vocabulary) defines procedure as: "A specified way to perform an activity." These procedures provide direction on how the company's quality policy is implemented. They explain in detail how each of the International Standard requirements is satisfied, and it is here the what, when, where, and who questions are answered. These procedures may contain proprietary information. Preparing this documentation is usually the responsibility of individual departments. The auditors' initial focus is on this level of documentation as they make up the nuts and bolts of the operations.

Structure of Level 2 Procedure

A good level 2 procedure usually contains the following sections:

- **Purpose and Scope**

  Defines the why, what for, and areas impacted by the procedure.

- **Responsibility**

  States the organizational unit or functional title responsible to implement the document to achieve the stated purpose.

- **Procedure**

  Lists, step-by-step, the details of what needs to be done in a logical sequence.

- **Flowchart**

  Shows a graphical representation of the procedure.

- **References**

  Identify which supporting documents or forms are associated with this document.

- **Records**

  Identify which records are generated as a result of using this document and how they are maintained. This information can also reside in a separate procedure.

3.3 Work Instructions/Level 3

Work instructions provide direction on how the procedure is to be implemented or how its requirements are to be accomplished. Work instructions define how to make, inspect, maintain, protect, transport, store, and repair products. They are normally very detailed and apply to a specific task.

A good level three work instruction always includes the man, machine, material, method, and environment (4 Ms and E). It is here the technical details, forms, and data sheets are included. These instructions are usually prepared by engineers and supervisors, and should include operator input.

Note: Auditors mostly question the operators when auditing level three documents.

3.4 Records/Standards/Level 4

Level four documents consist of reference material such as specifications, standards, and records. The ISO standards and QS-9000 can be considered level 4 documents.

Table 8.24 illustrates the various levels of documentation and their linkage. The example is for the contract review requirement.

Table 8.24: Documentation Levels Example

Policy:	Contract Review — A flowchart procedure is in place to ensure that the requirements of the customer are fully understood, confirmed back to the customer, and mutually accepted in writing by all concerned prior to scheduling of the order. Exceptions require executive approval. Related procedures: Repeat order processing #323-184 Custom order processing #323-185 Changes/amendments #323-186
Procedure:	Procedure #323-184 — Orders against existing contracts are received from customers at New York, Houston, or St. Louis locations by telephone, fax link, or e-mail by the customer representative and are routed to the designated service representative for processing. Further processing shall follow a work instructions developed for each customer.
Work Instruction:	W.I. #800-123 — Order Entry Process for Jones Account: Enter Jones Account Number on order entry terminal, follow menu displayed. Check for customer special instructions on menu file. Enter proper special instructions code into data base if instructions provided by the customer agree with file information. Order entry process must respond to all menu questions. Reserve space in production schedule and fax or e-mail request for conformation to customer. When customer acknowledges, book the order and confirm back to customer with order number.

4 Who Prepares Quality System Documentation?

Traditionally, the quality department prepared procedures and provided them to other departments for implementation. After all, the quality department was responsible for all quality problems and issues.

This approach resulted in the development of an ineffective quality system, as the communication was one way. Many non-conformances came from the fact that the departments had learned to develop the best procedures using their own methods, but these methods did not reflect those documented.

Today, documentation preparation should be a shared responsibility, with department personnel documenting their respective area requirements as explained by the "Document Owner and User Partnership." Only the individuals closest to the task know the procedures, and that is where the documenting should start. The standard "experts" and quality departments will be best utilized as resources to review documents, ensuring all parts of the standard are addressed.

4.1 Document Owner and User Partnership

4.1.1 Document Owner

The document owner is the department responsible for documenting the requirement. The document owner is also responsible for the implementation and upkeep of the documented requirement.

4.1.2 Document User

The document users are the departments affected and responsible for using the document prepared. Document owners are usually the biggest users.

The document owner and user partnership (Table 8.25) ensures the sharing of Quality System Documentation preparation by dividing the documentation requirements among the owners. Benefits of this approach are:

- Company-wide involvement
- Faster project completion
- Development of an effective system
- Documentation kept current

Table 8.25: Quality System Requirement Owners and Users

		TOP MANAGEMENT	QUALITY	ENGINEERING	OPERATIONS	MATERIALS	SALES	H.R. (OTHERS)
4.1	Management Responsibility	O	U	U	U	U	U	U
4.2	Quality System	U	O	U	U	U	U	U
4.3	Contract Review			U		U	O	
4.4	Design Control		U	O	U	U		
4.5	Document and Data Control	U	U	O	U	U	U	U
4.6	Purchasing		U	U	U	O		
4.7	Control of Customer-Supplied Product		U		U	O		
4.8	Product Identification and Traceability		U	U	O	U	U	
4.9	Process Control		U	U	O			
4.10	Inspection and Testing		O	U	U			
4.11	Control of Inspection, Measuring, and Test Equip.		U	O	U			
4.12	Inspection and Test Status		O	U	U	U	U	
4.13	Control of Non-Conforming Product		O	U	U	U	U	
4.14	Corrective and Preventive Action		O	U	U	U	U	
4.15	Handling, Storage, Packaging, Preserv., and Delivery		U	U	O	U		
4.16	Control of Quality Records	U	O	U	U	U	U	U
4.17	Internal Quality Audits	U	O	U	U	U	U	U
4.18	Training	U	U	U	U	U	U	O
4.19	Servicing				O	U	U	
4.20	Statistical Techniques		U	O	U			
2.1	Production Part Approval Process		U	U	U	U	O	
2.2	Continuous Improvements	O	U	U	U	U	U	U
2.3	Manufacturing Capabilities		U	O	U	U		

Note: These are typical examples, and your organization may be structured differently.

4.2 Document and Data Control Procedure Example

Figure 8.2 is a flowchart of a typical QS-9000 document and data control procedure.

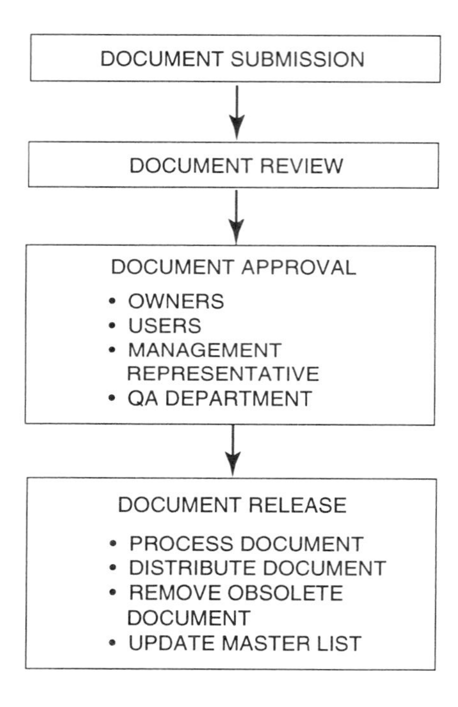

Figure 8.2 Document and Data Control Flowchart

Document Submission

Any employee can request to change and release a document by submitting a draft and Document Release/Change Form (Figure 8.3) to the document owner or a centralized document control group.

Document Review

This step involves a review between the originator, document owner, and user for consensus.

DOCUMENT RELEASE/CHANGE FORM

Document Change Order: ______________ Request Date: ______________

Document Number: ______________ Implementation Date: ______________

Document Title: ______________

Originator: ______________ Department: ______________

Document Owner: ______________

Reason for Change Request: ______________

____

Description of Change: ______________

____

____

____

Owner/User Approvals:

Dept.	Signature	Distribution	Dept.	Signature	Distribution

Management Representative Approval: ______________

Q.A. Approval: ______________

Revision History:

Rev.	DCO #	Date	Change Description

Figure 8.3 Document Release/Change Form

Document Approval

After agreement is obtained, the document is formally circulated for approval to the document approval board (owner, user, management representative, and QA). The minimum approvers should be stipulated in the procedures and readily apparent by looking at the change order documentation.

Document Release

After approval, the appropriate document control representative processes the document, distributes the document to users, retrieves obsolete documents, and updates the master list. One of the critical aspects of document control is ensuring that only the current revisions of documents and data are in use. Obsolete forms are the worst offenders and require special attention as copies are often unquantified.

5 How to Develop Quality System Documentation

Developing quality system documentation is a two-step approach:

- Design the quality system
- Document the quality system

5.1 Design the Quality System

The design phase is a key activity in the development of quality system documentation. A properly designed system will require fewer modifications later.

The project team is usually responsible for quality system design and the rules are very similar to those used for designing a product.

A good way of designing a quality system is to map or flowchart all activities starting from the customer inquiry through sales, manufacturing, final inspection, and delivery to the customer. Figure 8.4 is an example of how the flowcharting process works for a QS-9000 organization.

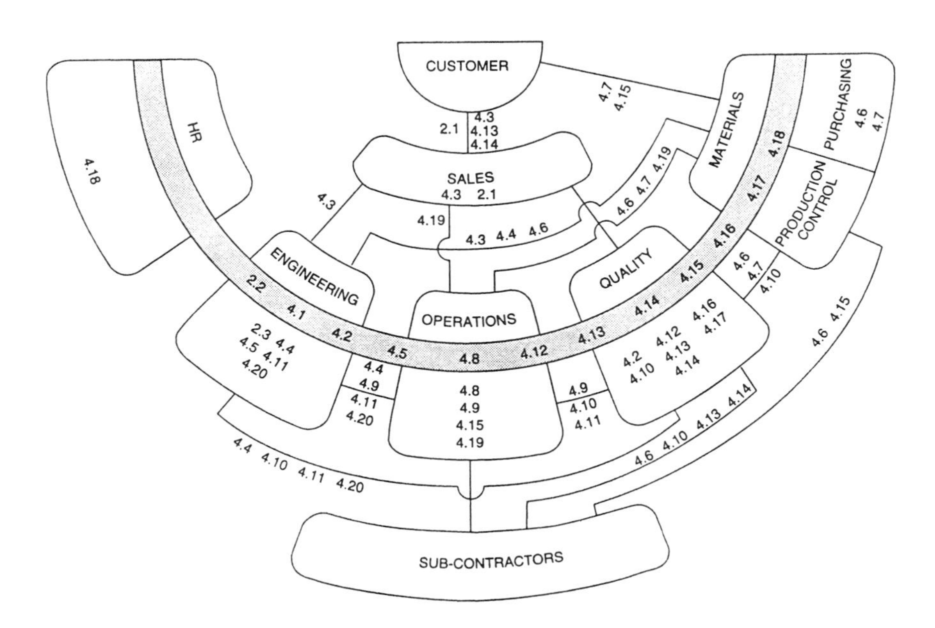

Figure 8.4 System Activity Flowchart

Each department (box) contains QS-9000 requirement numbers for which the department is responsible: requirements in the shaded area represent typical company-wide requirements, and the requirements listed on the line indicate the link between the various departments.

Here is a simplified explanation of how the flowchart works.

- **Customer**

  Sales receives customer requirements per 4.3 (Contract Review).

- **Sales**

  Sales reviews the requirements with engineering and materials per 4.3 (Contract Review). Special attention should be given to customer specific requirements in QS-9000 Section III. If satisfactory, the order is accepted. Note the linkage between sales and engineering.

- **Engineering**

  Engineering provides bill of materials and specifications to materials utilizing 4.3 (Contract Review). APQP and control plan processes come into play.

- **Materials**

  Materials purchases components required from sub-contractors per 4.6 (Purchasing) and schedules the order with operations through production control.

- **Quality**

  Quality inspects purchased product per 4.10 (Inspection and Testing) prior to use by operations.

- **Operations**

  Operations produces product conforming to 4.9 (Process Control).

- **Quality**

  Upon completion the quality department inspects product per 4.10 (Inspection and Testing) and, if acceptable, releases product to materials.

- **Materials**

  Materials ships product to customer per 4.15 (Material Handling, Packaging, Preservation, and Delivery).

5.2 Document the Quality System

Once the quality system has been designed, it must be documented. A documentation "overhaul plan" is shown in Figure 8.5.

Identify QS-9000 Documentation Requirements

Every quality system requirement mandates the preparation of documented procedures.

Collect and Review Existing Documentation

Collect all existing quality system documentation and review it for compliance to the applicable standard requirements. A checksheet (Figure 8.6) or the Quality System Assessment (QSA) can be used in this documentation review process.

- The QS-9000 requirements are listed on the checksheet under the "QS-9000 Requirement" column.
- Review each existing document and place the document number in the "Doc. #" column corresponding to the requirement it meets.
- If not already decided, it is wise to determine the proposed "Owners" and "Users" at this time. This information will speed up the registration effort.
- Any explanations should be written in the "Comments" column.

Note: This checksheet is a written record and will be useful be to the document owner during the "overhaul" process. This sheet can also be used as a reference during the various audits.

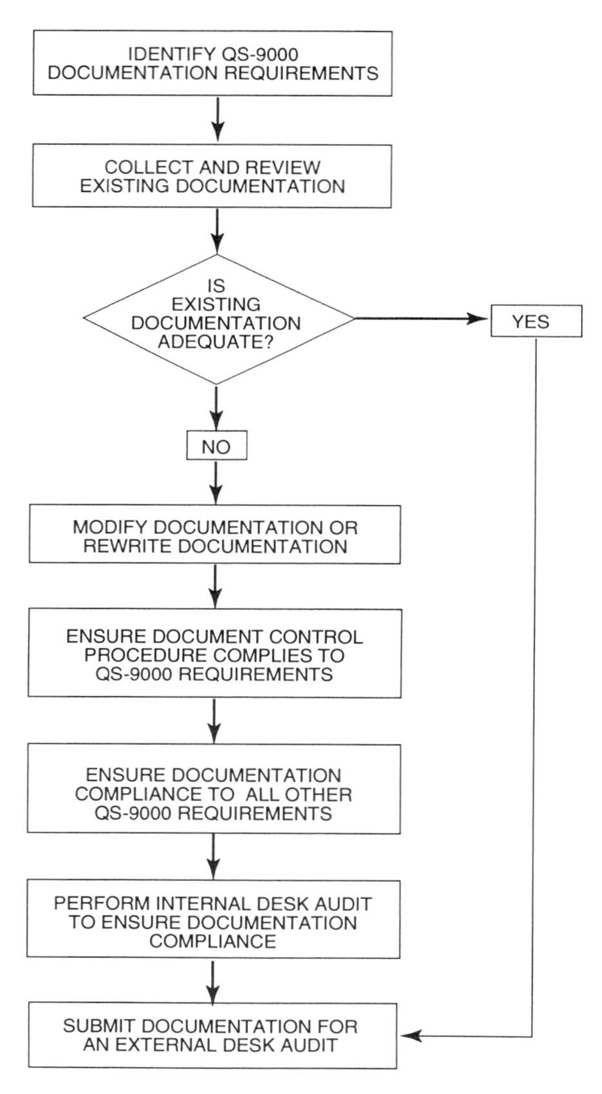

Figure 8.5 Documentation Overhaul Plan

QS-9000 Requirement	Doc. #	Owner	User	Comments
4.1 Management Responsibility	QSD-1A	Top Management	All	
4.2 Quality System	QSD-2B	QA	All	
4.3 Contract Review				No documents available

Figure 8.6 Documentation Review Checksheet

Is Existing Documentation Adequate?

The document review process leads to one of two conclusions:

Yes – Existing documentation is acceptable.

In such a scenario the organization can submit documentation to the registrar for the desk audit.

No – Existing documentation is not acceptable.

If the documentation is not acceptable, there are two options available:

1-Modify documentation

Existing documentation complies partially to requirements and minor documentation modifications will bring documentation into compliance.

2-Rewrite documentation

Existing documentation is inadequate and needs to be rewritten to comply to the documentation requirements. Start from scratch, flow the process, identify all related documents and records, and document the process.

Ensure Document Control Procedure Complies to QS-9000 Requirements

The document control procedure must be compliant with the QS-9000 requirements first as all other quality system procedures will be issued through it. This procedure should be developed by the project team; however, the ownership to maintain the document control system can be assigned to any department. In most companies document control is associated with the quality or engineering departments.

Note: It is possible to have multiple document control procedures for: policies, procedures, process work instructions, engineering changes/tests, drawing control, etc. Most differences are in the levels of document approval.

Ensure Documentation Compliance to Other QS-9000 Requirements

Company-Wide Requirements

These requirements are used by everyone in the company and should be documented first. Listed below are general company-wide requirements:

- Management responsibility
- Quality system
- Document and data control
- Product identification and traceability
- Inspection and test status
- Control of non-conforming product
- Corrective and preventive action
- Handling, storage, packaging, preservation, and delivery
- Control of quality records
- Internal quality audits

- Training
- Continuous improvement

Area-Specific Requirements

These are requirements not used by everyone in the company. Listed below are area-specific requirements:

- Contract review
- Design control
- Purchasing
- Control of customer-supplied products
- Process control
- Inspection and testing
- Control of inspection, measuring, and test equipment
- Servicing
- Statistical techniques
- Production part approval process (PPAP)
- Manufacturing capabilities

Perform Internal Desk Audit to Ensure Documentation Compliance

Upon the completion of quality system documentation, the project team conducts an internal desk audit to ensure that compliance to the requirements of QS-9000 has been achieved.

Any issues or non-conformances are resolved at this time.

Submit Documentation for External Desk Audit

When the project team is satisfied with the quality system documentation, the appropriate documentation is submitted to the registrar for an external desk audit.

5.3 How to Implement Quality System Documentation

If designing the QSD was "document what you do," then implementing QSD is "doing what you document." Believe it or not, implementation can be fun! Building the airplane is not half the fun of flying the airplane.

Implementation also has to be carefully planned. Implement the document and data control documentation first. For other documentation there is no one best plan. Some companies choose to implement individual procedures upon their completion, some implement all together, while others start at the inputs (contract review, purchasing) to the system and work their way into the system in a logical progression.

There should be no surprises since every owner and user is thoroughly familiar with the documentation. If there are some problems, don't panic; that's where continuous improvement comes in.

It may be a good idea to perform a trial implementation on a limited scale. Remember that the organization will not be ready for either a "desk audit" or a "pre–registration audit" without complete documentation of the quality system. Also remember that it is not sufficient to have the documentation completed, these documents should be seasoned by use. Some documentation such as corrective action must be in use long enough to have collected a history of corrective actions through incident to closure.

6 Conclusion

As each of the requirements of the QS-9000 becomes understood, the components of the quality system, together with the appropriate documentation will, like bricks in a structure, fall into place.

It is important to remember that development and implementation of QSD is not an event, but an ongoing process and is to be continually improved using TAP-PDSA. Always keep in focus that the primary purpose of a quality system is to ensure customer satisfaction and profitability, not just QS-9000 certification.

Chapter 9
On the Road Again

After registration, the company has an opportunity to look back at what has been accomplished and to look forward to what the future holds.

- Was QS-9000 registration worth the effort?
- What's next?

1 Was QS-9000 Registration Worth the Effort?

To answer this question, it is necessary to go back to the reason why the registration project was initiated in the first place, and to verify that the objectives were met. This is the study step in the TAP-PDSA.

If the reason was:

- **Customer Requirements**

Are the customers satisfied? Are there signs of improved customer relations? Did the company receive any customer awards?

- **Marketing Strategy**

Was there an increase in sales? Has the market share gone up? Did the company acquire new customers as a result of registration?

If the answers to the above questions were less than an unqualified yes, then the company needs to continue the TAP-PDSA cycles to meet the objectives by taking the appropriate corrective actions.

2 What's Next?

QS-9000 registration is only an initial requirement. The company should now plan to take the quality system to the "next level" as explained by the following:

- Sustain registration through continuous improvement.
- Evaluate the global impact on the quality system.
- Prepare for future generation requirements.
- Revisit Dr. Deming's 14 points of management.

2.1 Sustain Registration Through Continuous Improvement

After the company is registered, it must sustain registration through a system of periodic surveillance audits. Experience has shown that many companies "take it easy" after becoming registered and the first surveillance audit comes as a rude awakening that the system is not performing as effectively as it should.

This echoes Deming's first point of management: "constancy of purpose for continuous improvement." Maintain QS-9000 registration diligently and build upon it. QS-9000 registration is an excellent platform to scale the heights of quality excellence, not a signal that the quality project has been completed.

Constantly improve the system (4 Ms and E) to improve quality and delivery, reduce costs to stay ahead of competition.

Use TAP-PDSA to re-evaluate the goals and objectives of the system. Develop short-term and long-term goals. For example, select a few quality system requirements each quarter and take them to the next level of improvement (Table 9.1).

Table 9.1: Quality System Improvement

Quality System Requirement	Next Level Development
Management Responsibility	Management Measurement System
Quality System	Simplify System Structure
Document and Data Control	Paperless Document Control System
Purchasing	Supplier Management Team
Product Identification and Traceability	Bar Code System
Quality Records	Paperless Record Keeping System
Internal Quality Audits	Cross Functional Audit Teams
Training	Internal Training University
Statistical Techniques	Company-Wide SPC System

Example: Develop a supplier management team (materials, quality, engineering, operations) and drive quality requirements down to the sub-contractors. Move the quality system requirements onto the supplier base. "Ship to stock" has the potential for immense cost savings and quality improvements. Use the internal auditors to evaluate and assist suppliers to become QS-9000 compliant.

Similarly, select other quality system requirements over a period of time and improve them. This overall plan becomes a *strategic quality improvement plan* (SQP).

2.2 Evaluate the Global Impact on the Quality System

Prepare the quality system diligently for a global economy. Surveys indicate that 75% of companies that do not market globally will do so by the end of the decade or risk going out of business. Don't be late. Those companies planning to market products globally must be aware of the European Union (EU) and the product registration mark CE.

2.2.1 What Is the European Union? and What Does It Do?

The European Union (EU) is a treaty organization. It originated with the 1957 Treaty of Rome. One of its objectives was to abolish tariffs and trade quotas among the member countries and to stimulate economic growth within the member states. The EU acts as a sovereign legal entity controlling the activities assigned to it by the treaty. An early target for the EU was defining national product certification requirements. Differences in product certification requirements made selling products in the various countries a difficult undertaking because of the need for duplication of conformity tests, certification, documentation, and the requirement for separate approvals, often from both national and local regulatory agencies.

The Treaty of Rome was amended in 1986 to incorporate the "Single European Act," which established the principle that products that meet the requirements of one EU member state could freely circulate in other member states. This is a concept which is very similar to the Interstate Commerce article in the U.S. constitution. All of this single market activity became known as EC 92 and was scheduled for implementation at the end of the 1992 calendar year. The target year was found to be unrealistic and the goal is presently being implemented progressively.

2.2.2 The CE Mark

Quality systems must become increasingly sophisticated to accommodate the more stringent product quality requirements that are coming out of the European Union and North America (particularly the United States and Canada). Go into any electronics store and inspect the back of the base on a piece of equipment. On many pieces you will find these approval labels: UL, CSA, TUV/GS, BZT, and the CE mark.

The CE mark is the most progressive.

What Is a CE Mark?

The CE mark is a product registration mark which identifies that the product was manufactured under EU normalization directives. The European Union has adopted a number of directives that require certain products entering its market to bear a CE mark and appropriate technical documentation indicating compliance to the requirements of the directive.

The following are some directives:

89/336/EEC - The Machinery Directive. This directive covers all machinery, defined as "an assembly of linked parts or components, at least one of which moves, with the appropriate actuators, control and power circuits, etc., joined together for a specific application, in particular for the processing, treatment, moving or packaging of a material." The directive outlines the essential safety requirements for this equipment. It was adopted on June 14, 1989, and went into effect on January 1, 1995.

93/42/EEC - The Medical Devices Directive. This directive covers medical devices and their accessories. For the purposes of this directive, accessories shall be treated as medical devices in their own right. Definitions of "medical devices" can be found in the article 1 definitions, scope 2 (a) of this directive. This directive outlines the scope, classifications, conformity assessment procedures and essential requirements. It was adopted on June 14, 1993, and goes into effect on June 14, 1998.

Products under the scope of directives will not be allowed to be sold in the European Market unless they bear the appropriate CE mark and technical documentation.

What Is the Purpose of the CE Mark?

The purpose of the CE Mark is to provide for the "free movement" of goods across national boundaries. The EU established legal requirements known as directives, which are intended to harmonize the existing legal practices in the member states. The national governments of the member states are obligated to harmonize national law with the directives. Any national laws that conflict with the directives are required to be repealed.

How Does an Organization Obtain a CE Mark for Its Product?

The CE mark can be obtained through self-declaration or through the use of a notified body:

- Self-Declaration (declaration of conformity)

The manufacturer performs tests conforming to the appropriate EU directive and prepares a test report indicating that the product complies with the appropriate directive. Upon compliance, the manufacturer affixes a CE mark to the product.

- Use of a Notified Body

For certain products, a notified body must be used to perform or verify the required tests and upon compliance provide a certificate of conformity. The manufacturer is then allowed to affix the product with a CE mark.

Notified Body: A notified body is an agency considered competent in the field of a specific directive and who is authorized by EU member states to test, verify, and certify product per the directive.

Note: Some EU directives may require ISO 9000 registration as a condition to obtaining the CE mark.

2.3 Prepare for Future Generation Requirements

There are several standards under development today which will become the requirements of tomorrow. To stay ahead, evaluate their impact on the company's business plan.

- ISO 14000, Environmental Management System Standards
- Future ISO 9000 Standards (1999)

2.3.1 ISO 14000

What Are the ISO 14000 Series of Standards?

In 1993, ISO created TC-207 to develop environmental standards, i.e., the ISO 14000 Series of Standards. Table 9.2 lists the ISO 14000 series of standards under development.

Table 9.2: ISO 14000 Series of Standards

ISO 14001	Environmental Management System Specification (analogous to ISO 9001)
ISO 14004	Environmental Management System–General Guidelines on Principles, Systems, and Supporting Techniques
ISO 14010	Guidelines for Environmental Auditing–General Principles of Environmental Auditing
ISO 14011	Guidelines for Environmental Auditing–Audit Procedures-Part I: Auditing of Environmental Management Systems
ISO 14012	Guidelines for Environmental Auditing–Qualification Criteria for Environmental Auditors

What Is the Purpose of ISO 14001?

ISO 14001 specifies the requirements of an environmental management system.

Compliance to ISO 14001 will demonstrate an organization's ability to meet environmental objectives and targets. ISO 14001 is generic in nature, and it will therefore be applicable to any type of organization.

2.3.2 Future ISO 9000 Standards (1999)

The ISO 9000 series of standards were revised in 1994. The next major revisions are expected in the year 1999. Based on the outcome of the TC 176 annual meeting in Durban, South Africa, the following are insights into the future of ISO 9000 and its series of standards:

ISO 9001 Consolidation

ISO 9001, 9002, and 9003 may be consolidated into a single standard ISO 9001 to be published around the year 1999. Companies currently registered to ISO 9002 will then be registered to ISO 9001 without design control, but for companies currently registered to ISO 9003, the upgrade is expected to involve additional requirements.

ISO 9001 Reorganization

Speakers at the ISO 9000 forum symposium looked at the future architecture of ISO 9001. A concept plan showed that the twenty clauses had been consolidated into the following major groupings:

- Executive Management
- Process Management
- Measurement
- Evaluation and Improvement

ISO 9000 Consolidation

ISO 9000 document may consolidate 9000-1,2 etc., and become a more comprehensive document. ISO 9000 is also projected to include an expanded vocabulary and technical terms, presently ISO 8402.

2.4 Revisit Dr. Deming's 14 Points of Management

Chapter 3 explained the importance of the 14 points of management as they related to the QS-9000 project. In this chapter we look at each of these points in detail as they apply to an organization's pursuit for continuous improvement (as explained in the video "The 14 Points of Management," courtesy MIT Center for Advanced Engineering Study).

Note: In many instances the words of Dr. Deming have been paraphrased to meet the intent of this book.

Dr. Deming: These 14 points provide management a roadmap for transforming any organization (manufacturing or service) into a quality-focused, customer-driven business able to capture and compete in world-wide markets. The fourteen points are the obligation of top management to work on forever.

Point #1: Constancy of Purpose

Top management must establish constancy of purpose for improvement of product and service. This means developing a goal statement that includes the organization's objectives and guidance for achieving them. Guidance plan should provide for both short-term and long-term operations. Establishment of constancy of purpose includes putting resources into innovation, research, training, and the constant improvement of product and service.

Dr. Deming: Innovation will be required: allocate resources for long-term planning, plans for the future, or for consideration of ideas and tests. Look ahead for new services and new products that may help people to live better materially and which will have a new market. Take into consideration the performance in the hands of the customer, how will he use, misuse, etc. Until this policy of constancy of purpose can be enthroned as an institution, middle management, top management, and everyone else in the company will be skeptical. It is necessary to put resources into research and education. The consumer is the important end to the production line.

If you don't have a consumer, someone to buy your product or use your service, you might as well shut down and close up.

Point #2: Everybody Wins

Dr. Deming: Adopt a new philosophy. We can no longer accept defects, common levels of mistakes, everything late as an accepted way of life. Think of the economy that would come from dependable products and services. Reliable services reduce cost, defects raise cost. Think of the cost of living with waste in it. How much could you get with the same dollar if there was no waste embedded in it?

Point #3: Design Quality In

Depending on mass inspection to assure quality output is a snare and delusion. Management must understand the purpose of inspection, which is for improvement of process and reduction of cost.

Dr. Deming: Inspection does not improve quality, the quality is already there. Inspection raises the cost. It only indicates that somebody doesn't know how to do the job.

Quality – good or bad – is built into products or services. High quality results from robust design, statistical process control during manufacturing, and a total management system that focuses on innovation and a commitment to the future. Inspection to improve quality is too late, too costly, and ineffective. Defects are not free; somebody makes them and gets paid for making them. High quality comes not from inspection, but from the improvement of the process. Inspection, scrap, downgrading, and rework are not corrective action on the process.

Point #4: Don't Buy on Price Tag Alone

End the practice of awarding business to the lowest bidder without regard to quality.

Dr. Deming: Quality has to come into consideration; there is no such thing as price without knowledge of quality. What buyer is able to judge quality? He can judge price, but to learn something about quality is a five-year program and there is some learning to do. He has to learn process control by statistical methods, and learn that it is only by statistical evidence that he can know what a vendor is able to produce. We must buy materials whose quality is demonstrable before it leaves the door of the vendor. That demonstration can only come by statistical charts to be furnished by the vendor.

With evidence of quality, the purchasing department must change its focus from lowest initial cost to lowest total cost. Total cost includes the purchase price, plus the costs to correct problems encountered in production and in use.

Dr. Deming: We must do business with vendors that can furnish evidence of quality, which only means evidence that they know what they are doing, that they know their cost, that they can produce the same stuff tomorrow and next week, and at the same or lower cost. They must continually improve. This means intervention by management with the system, with the aim of improvement. This will mean in many instances that companies will find only one vendor which is qualified. The problem is to find one: "in order to find two you have to find one first." A company must adopt a philosophy to work with the vendor and his people; this becomes a chain reaction. As you help your vendor to improve, you help the whole industry to improve.

Point #5: Continuous Improvement

To remain competitive, improve constantly and forever the system of production and service. The PDSA cycle provides an overview of the process of continuous improvement. Use the cycle again and again and the process of continual improvement carries on.

Point #6: Training for Skills

Institute a vigorous training program. Use statistical methods in training to determine its effectiveness. Training should not be limited and restricted. Training is needed anytime: when an employee is new to a job, when responsibilities shift, and when new equipment or procedures are added. Train employees the right way and get them in statistical control.

Point #7: Institute Leadership

The aim of supervision should be to help employees do a better job. Supervision should not be criticism, policing, or searching for individual wrong doing. Workers should not be blamed for problems that are beyond their control to prevent or correct. Management must change from a "who's wrong" to a "what's wrong" philosophy. Ineffective or misguided supervision is time consuming and all too often demoralizing.

Supervisors need to learn how to lead people in the process of continual improvement. This will only happen when the supervisors themselves are properly trained. They must learn how to use statistical tools to find and correct the sources of defectives within the system. Top management must support their efforts by creating an environment that fosters improvement of the system through effective leadership and supervision.

Point #8: Drive Out Fear

Dr. Deming: Employees are afraid to ask questions. Many of them are afraid to take a position. Fear prevents employees from serving the best interests of the organization. The economic loss from fear is appalling. To achieve better quality and productivity, it is necessary to drive out fear and to make people feel secure. Secure means without concern or worry, a feeling of being able to communicate with people, to offer suggestions and criticism, and to report problems.

Point #9: Break Down Barriers

Dr. Deming: Break down barriers between employees. Why can't people talk with each another? Engineering, design, production, and marketing should work together, learn from each other, and pool knowledge to achieve common goals. Only top management can bring people together and reduce fear.

Point #10: Eliminate Slogans

Slogans, posters, and exhortations asking workers to improve quality and productivity do nothing to help employees meet these goals.

Dr. Deming: Employees need to know how. Management must provide methods for achieving goals. The "how" must start at the top. If management does not carry out the 14 points, the transformation won't happen.

Point #11: Method

A 5% allowance for defects, 10% allowance for scrap won't help anyone do a better job. Work standards and quotas tell employees what management wants, but do not help them to achieve these goals. This is demoralizing and stifles pride of workmanship. Often, the only way to make the quota is to knowingly produce defectives. When people are asked to make quotas, quality usually suffers.

Point #12: Joy in Work

There are many barriers to pride of workmanship that prevent people from working effectively within the organization (e.g., defective incoming materials, poor supervision, inadequate training, posters, slogans, and quotas). Employees, both salaried and hourly, feel a loss of pride when they know their product or service is shoddy, poorly designed, unsafe, or defective. Another barrier to pride of workmanship is the feeling among many workers that they are being treated as commodities. They work long and hard hours for the organization, only to be laid off when profits start to slip. How can employees feel loyalty and pride in an organization that fails to recognize their contributions and needs? How can they contribute their best efforts today, when they don't know if they will have a job tomorrow? Loss of pride due to anxiety and fear is not uncommon. Once these barriers to pride of workmanship are removed, the organization can reap the benefits of employees that are able to work to their maximum potential.

Point #13: Continuing Education

To remain competitive, institute aggressive programs for education and self-improvement. Education in simple but powerful statistical techniques is necessary for everyone in top management. With everyone speaking the same language, it becomes easier to communicate objectively of problems, solutions, and improvements to the system. Education and self-improvement will help employees improve their work and serve the organization more effectively. Support for degree programs will assist employees to advance in their careers and keep pace with the organization as it grows and as technology changes. Job-related subjects, such as electronics and computers skills, will improve efficiency and performance. Programs aimed at personal self-improvement will also benefit the organization's bottom line. Courses and facilities for improving health, stopping substance abuse, reducing stress, and preventing burnout will save more money in the long term and substantively improve morale and performance. Developing the human potential through employee education and self-improvement reinforces the organization's commitment to long-term growth and survival.

Point #14: Accomplish the Transformation

Dr. Deming: Organize to move everyday, not just once in a while, on the 14 points. Take actions to accomplish the transformation. Top management must provide this direction and guidance to make these changes. Review on a regular basis the progress on each of these 14 points. Make things better and lead the way. Allow adequate time for the transformation. Don't be fooled by quick success.

In conclusion Dr. Deming states: Tangible results from the 14 points won't all come at once; some will come quicker than others, and the situation will take care of itself in nature's way. Companies that adopt constancy of purpose, establish it as an institution to improve quality and productivity, and go about it with intelligence and perseverance have a chance to survive.

3 Conclusion

Changes to the standards are going to be more dramatic in the next 5 years than they have been in the last 25 years. The global market is forcing the globalization of standards. Competition has become fiercer, and to stay in business companies must continually improve and at the same time comply with the stringent emerging requirements.

Train and educate continuously to learn about future trends and requirements. Perform an ***analysis*** of the existing situation and then predict a ***plan*** with the highest probability of success. Finally, implement and "***lead the way***."

Bibliography

Beaumont, Leland R. *ISO 9001, The Standard Companion* (2nd ed). Middletown, NJ: ISO Easy, 1993

Fellers, Gary. *The Deming Vision: TQM for Administrators.* Milwaukee, WI: ASQC Quality Press, 1992

Clements, Richard Barrett. *Quality Manager's Complete Guide to ISO 9000.* Englewood Cliffs, NJ: Prentice Hall, 1993

Crosby, Phillip B. *Quality Is Free, The Art of Making Certain.* New York, NY: McGraw-Hill, 1979

Delavigne, Kenneth T., and Robertson, J. Daniel. *Deming's Profound Changes.* Englewood Cliffs, NJ: PTR Prentice Hall, 1994

Deming, W. Edwards. *Out Of Crisis.* Cambridge, MA: MIT Center for Advanced Engineering Study, 1986

Deming, W. Edwards. *The New Economics for Industry, Government and Education.* Cambridge, MA: MIT Center for Advanced Engineering Study, 1993

Gluckman, Perry, and Roome, Diana Reynolds. *Everyday Heroes, from Taylor to Deming; The Journey to Higher Productivity.* Knoxville, TN: SPC Publications, 1990

International Organization for Standardization. *ISO 9000 International Standards for Quality Management.* Geneva, ISO, 1994

Ishakawa, Kaoru. *Guide to Quality Control* (2nd ed.). Tokyo: Asian Productivity Organization (in the United States, UNIPUB, New York, NY) 1985

Johnson, L. Marvin. *Quality Assurance Program Evaluation* (revised edition). Los Angeles, CA: L. Marvin Jones, Associates, 1990

Juran, J. M. *Management Breakthrough.* New York, NY: McGraw-Hill, 1964

Kivenko, Kenneth. *Quality Control for Management.* Englewood Cliffs, NJ: Prentice Hall, 1984

Lamprecht, James L. *ISO 9000: Preparing for Registration.* Milwaukee, WI: ASQC Quality Press, 1992

Lamprecht, James L. *Implementing the ISO 9000 Series.* New York: Marcel Dekker and Milwaukee, WI: ASQC Quality Press, 1993

Mills, Charles A. *The Quality Audit, A Management Education Tool.* New York, NY: McGraw-Hill, 1989

Peach, Robert W. (ed.). *The ISO 9000 Handbook* (2nd ed.). Fairfax, VA: CEEM Information Services, 1994

Scherkenbach, William W. *The Deming Route to Quality and Productivity: Roadmaps and Roadblocks.* Washington, DC: CEEP Press Books, 1986

Senge, Peter. *The Fifth Discipline, The Art and Practice of the Learning Organization.* New York, NY: Doubleday/Currency, 1990

Walton, Mary. *The Deming Management Method.* New York, NY: The Putnam Publishing Group (Petigree), 1986

Index

A

B

C

D

G

I

L

M

N

P

Q

R

S

T

W